PSYCHOPHARMACOLOGY

STEPHEN M STAHL MD PhD

Director
Clinical Neuroscience Research Center
and
Adjunct Professor of Psychiatry
University of California
San Diego

*Donated as a service
to mental health by*

MARTIN DUNITZ

Although every effort has been made to ensure that the drug doses and other information are presented accurately in this publication, the ultimate responsibility rests with the prescribing physician. Neither the publishers nor the author can be held responsible for errors or for any consequences arising from the use of information contained herein.

© Stephen M Stahl 1999

First published in the United Kingdom in 1999 by
Martin Dunitz Ltd, 7–9 Pratt Street, London NW1 0AE

ISBN 1-85317-601-X

Composition by Scribe Design, Gillingham, Kent, UK
Printed and bound in the United States of America

CONTENTS

ACKNOWLEDGEMENT

Figures in *Psychopharmacology of Antipsychotics* are reproduced with the permission of Cambridge University Press from Stahl, S.M., *Essential Psychopharmacology* (Cambridge University Press, 1996), and from additional figures copyright Stephen M. Stahl, 1996 and 1999.

CHAPTER 1

INTRODUCTION

This pocketbook is a series of visual lessons on the antipsychotic drugs. The more than one dozen conventional antipsychotics, sometimes also called 'classical neuroleptic agents' are now being replaced in clinical practice by the atypical antipsychotics, which include about half a dozen marketed agents, with several more in the late stages of clinical development.

To understand the mechanism of action of the new atypical antipsychotics as well as the conventional antipsychotics, it is necessary to grasp several pharmacologic concepts relating to four key neurotransmitter systems: especially dopamine and serotonin, but also acetylcholine and glutamate. These are covered in Chapter 2. To appreciate the breadth of clinical applications for the new atypical antipsychotics, it is important to review the evolving concepts about the multiple symptom dimensions in schizophrenia. These include not only positive and negative symptoms, but also cognition, aggression and hostility as well as depression and anxiety in schizophrenia. These are discussed in Chapter 3.

During the 1950s and 1960s, the conventional antipsychotics revolutionized the treatment of schizophrenia, leading to the closing of hospitals for most chronic psychiatric patients. These drugs dominated the treatment of schizophrenia and other psychoses for over 30 years. Although these agents are now being replaced by atypical antipsychotics, they are still in significant use in some countries. These conventional antipsychotic drugs, also known as the classical neuroleptic agents, are discussed in Chapter 4.

During the 1970s and 1980s, a few additional antipsychotic agents were marketed from the conventional antipsychotic class, but it was not until the re-discovery of clozapine in the late 1980s that the atypical antipsychotics were born. In the past few years, several new agents have again revolutionized the treatment of schizophrenia in the 1990s. These newer agents have all exploited the dual action of clozapine upon serotonin as well as dopamine.

This pharmacologic dimension of the atypical antipsychotics is reviewed in detail in Chapter 5, which points out the three potential advantages of having dual serotonin-dopamine action rather than dopamine action alone: namely, reduction of extrapyramidal symptoms and tardive dyskinesia; reduced incidence of hyperprolactinemia; and improved efficacy for treating the negative symptoms of schizophrenia.

Today, there are three atypical antipsychotic agents in addition to clozapine marketed in the USA: risperidone, olanzapine and quetiapine. In Europe and other markets, a fourth agent, sertindole, has recently been withdrawn from marketing. Several other agents are also in late phase clinical testing, such as ziprasidone. Each of these specific drugs is discussed in Chapters 6 to 8.

The atypical antipsychotics have a vast number of pharmacologic properties beyond serotonin-dopamine antagonism. Such properties may not only be responsible for the distinctions between atypical antipsychotics, but also be the reason why clinical research is beginning to uncover additional therapeutic uses for the atypical antipsychotics beyond the treatment of the positive and negative symptoms of schizophrenia. The dozen or so pharmacologic properties of the atypical antipsychotics are discussed in Chapter 6, the distinguishing pharmacokinetic properties of the atypical antipsychotics examined in Chapter 7, and the expanded clinical applications of the atypical antipsychotics are discussed in Chapter 8. They include use in the treatment of positive symptoms for disorders other than schizophrenia, as well as the treatment of cognitive impairment, aggression and depression in schizophrenia and other psychiatric disorders.

The figures and diagrams in this pocketbook are based largely on the textbook *Essential Psychopharmacology* (Cambridge University Press, 1996), which the reader is encouraged to consult for further details and references. Additional references are included at the end of this book.

CHAPTER 2

FOUR KEY NEUROTRANSMITTER SYSTEMS

This chapter presents the neurobiological basis of antipsychotic action by explaining important aspects of four neurotransmitter systems in the central nervous system (CNS). The four neurotransmitter systems are:

- dopamine (DA)
- serotonin (also 5-hydroxytryptamine or 5HT)
- acetylcholine (ACh)
- glutamate (Glu).

A series of figures shows how each neurotransmitter is synthesized and metabolized as well as how each interacts at various receptors. Finally, several diagrams explain the manner in which dopamine controls the release of acetylcholine, and how serotonin controls the release of dopamine.

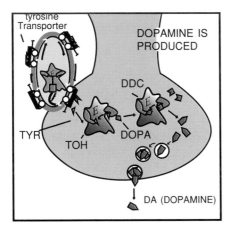

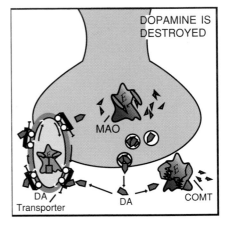

FIGURE 2.1. Dopamine (DA) is produced in dopaminergic neurons from the precursor tyrosine (Tyr), which is transported into the neuron by an active transport pump (tyrosine transporter), and then converted into DA by two out of three of the same enzymes (E) that also synthesize norepinephrine (NE, noradrenaline). The DA-synthesizing enzymes are tyrosine hydroxylase (TOH), which produces dihydroxyphenylalanine (DOPA), followed by DOPA decarboxylase (DDC), which produces DA.

For more information about dopamine see: Baldessarini 1995; Bond and Lader 1996; Civelli 1996; Cooper et al 1996; Gelenberg and Bassuk 1997; Grace and Bunney 1996; Hyman et al 1995; Janicak et al 1997; Le Moal 1996; Leonard 1997; Mansour and Watson 1996; Mansour et al 1998; Quitken et al 1998; Roth et al 1995; Roth and Elsworth 1996; Schatzberg et al 1997; Seeman 1995; Seeman 1996; Stahl 1996.

FIGURE 2.2. Dopamine (DA) is destroyed by the same enzymes (E) that destroy norepinephrine (NE, noradrenaline), namely monoamine oxidase (MAO) and catechol-O-methyl transferase (COMT). The DA neuron has a presynaptic transporter (DA transporter) which is unique for DA, but works analogously to the NE and serotonin (5-hydroxytryptamine, 5HT) transporters.

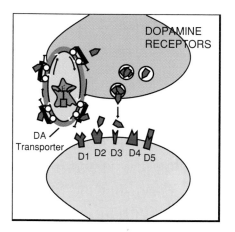

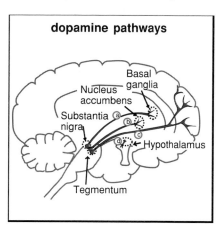

FIGURE 2.3. Receptors for dopamine (DA) regulate dopaminergic neurotransmission. A plethora of dopamine receptors exists, including at least five pharmacologic subtypes and several more molecular isoforms. Perhaps the most extensively investigated dopamine receptor is the dopamine 2 (D_2) receptor, because it is stimulated by dopaminergic agonists for the treatment of Parkinson's disease, and blocked by dopamine antagonists for the treatment of schizophrenia. This will be discussed in much greater detail later.

FIGURE 2.4. Four dopamine pathways in the brain. The neuroanatomy of dopamine neuronal pathways in the brain can explain both the therapeutic effects and the side effects of the known antipsychotic agents.

(a) The nigrostrial dopamine pathway projects from the substantia nigra to the basal ganglia, and is thought to control movements.

(b) The mesolimbic dopamine pathway projects from the midbrain ventral tegmental area to the nucleus accumbens, a part of the limbic system of the brain thought to be involved in many behaviors, such as pleasurable sensations, the powerful euphoria of drugs of abuse, as well as the delusions and hallucinations of psychosis.

(c) A pathway related to the mesolimbic dopamine pathway is the mesocortical dopamine pathway. It also projects from the midbrain ventral tegmental area, but sends its axons to the limbic cortex, where it may have a role in mediating positive and negative psychotic symptoms or cognitive side effects of neuroleptic antipsychotic medications.

(d) The fourth dopamine pathway of interest is the one that controls prolactin secretion, called the tuberoinfundibular dopamine pathway. It projects from the hypothalamus to the anterior pituitary gland.

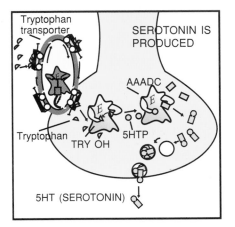

FIGURE 2.5. Serotonin (5-hydroxytryptamine, 5HT) is produced from enzymes after the amino acid precursor tryptophan is transported into the serotonin neuron. The tryptophan transport pump is distinct from the serotonin transporter (see Figure 2.6). Once transported into the serotonin neuron, tryptophan is converted into 5-hydroxytryptophan (5HTP) by the enzyme tryptophan hydroxylase (TRY-OH). 5HTP is then converted into 5HT by the enzyme aromatic amino acid decarboxylase (AAADC). Finally, serotonin is stored in synaptic vesicles where it stays until released by a neuronal impulse.

For more information about serotonin see: Aghajanian 1996; Azmitia and Whitaker-Azmitia 1996; Bond and Lader 1996; Cooper et al 1996; Gelenberg and Bassuk 1997; Glennon and Dukat 1996; Hyman et al 1995; Jacobs and Fornal 1996; Janicak et al 1997; Leonard 1997; Sanders-Bush and Canton 1996; Schatzberg et al 1997; Schatzberg and Nemeroff 1998; Shih et al 1996; Stahl 1996.

FIGURE 2.6. Serotonin (5HT) is destroyed by the enzyme monoamine oxidase (MAO), and converted into an inactive metabolite. The 5HT neuron has a presynaptic transport pump selective for serotonin called the serotonin transporter, which is analogous to the dopamine (DA) transporter in DA neurons (see Figure 2.2).

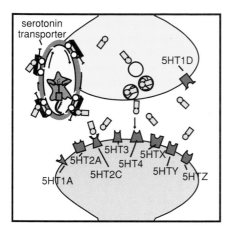

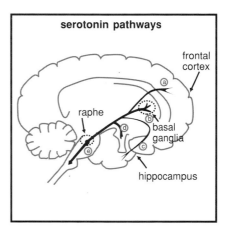

FIGURE 2.7. Receptor subtyping for the serotoninergic neuron has proceeded at a very rapid pace, with at least four major categories of serotonin (5HT) receptors, each further subtyped depending upon pharmacologic or molecular properties. In addition to the serotonin transporter, there is a key presynaptic serotonin receptor (the $5HT_{1D}$ receptor) and several postsynaptic serotonin receptors ($5HT_{1A}$, $5HT_{2A}$, $5HT_{2C}$, $5HT_3$, $5HT_4$ and many others denoted by $5HT_X$, $_Y$ and $_Z$).

FIGURE 2.8. Key serotonin pathways in the central nervous system. At least five pathways hypothetically serve as the substrate for the wide-ranging regulatory actions of serotonin. Antidepressant actions of drugs that inhibit the serotonin transporter are hypothesized to be mediated by the pathway from the midbrain raphe to the prefrontal cortex (pathway a). Other pathways to the prefrontal cortex may mediate cognitive effects of serotonin. The actions of drugs that treat obsessive-compulsive disorder (OCD) are hypothesized to be mediated by the pathway from the midbrain raphe to the basal ganglia (pathway b). This pathway theoretically mediates the regulatory actions of serotonin on movements as well. The pathway from the raphe to the limbic cortex may mediate serotonin's regulatory function upon emotions, including panic, memory and anxiety (pathway c). Effects of serotonin on eating behaviors and appetite may be mediated by the pathway from midbrain raphe to the hypothalamus (pathway d). Effects of serotonin upon sexual functioning may be mediated by the pathway that projects from the raphe down the spinal cord (pathway e). Still other centers in the brain stem mediate serotonin's control of the sleep-wake cycle.

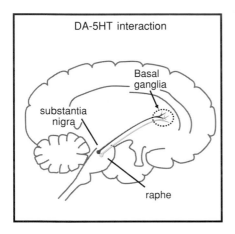

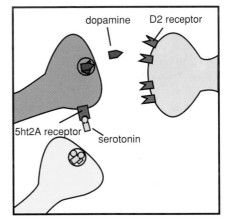

FIGURE 2.9. Serotonin-dopamine projections to the basal ganglia. Serotonin neurons from the midbrain raphe project to the basal ganglia (shown here as yellow neurons). Dopamine neurons from the substantia nigra in the brain stem also project to the basal ganglia (shown here as blue neurons). The axon terminals and the interactions between the serotonin and dopamine neurons there are shown in Figure 2.10.

For more information about serotonin-dopamine interactions see: Kapur 1996; Kapur and Remington 1996; Kinon and Lieberman 1996; Meltzer et al 1989; Stahl 1996.

FIGURE 2.10. Serotonin-dopamine interactions in the basal ganglia. Serotonin (5HT) in the yellow neurons normally has the ability to inhibit dopamine release in the blue neurons. This type of interaction may be via serotonin axons interacting with dopamine axons in an axo-axonal synapse. Alternatively, serotonin receptors on presynaptic dopamine axon terminals may be able to interact with serotonin which diffuses there from serotonin axons, but without a synapse. This serotonin interaction at dopamine neurons occurs via $5HT_2$ receptors.

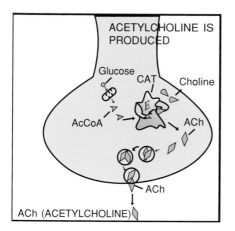

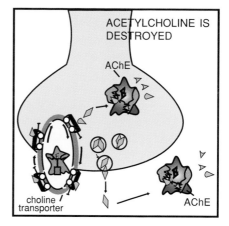

FIGURE 2.11. Acetylcholine (ACh) is produced. ACh is a prominent neurotransmitter, which is formed in cholinergic neurons from two precursors: choline and acetyl coenzyme A (Ac-CoA). Choline is derived from dietary and intraneuronal sources, and Ac-CoA is made from glucose in the mitochondria of the neuron. These two substrates interact with the synthetic enzyme choline acetyl transferase (CAT) to produce the neurotransmitter acetylcholine.

For more information on acetylcholine see: Bloom and Kupfer 1996; Cooper et al 1996; Schatzberg and Nemeroff 1998; Stahl 1996.

FIGURE 2.12. Acetylcholine (ACh) destruction and removal. ACh is destroyed by an enzyme called acetylcholinesterase (AChE), which converts ACh into inactive products. One of these is choline which can be pumped back into the neuron by a presynaptic choline transporter similar to the transporters for other neurotransmitters already discussed (dopamine and serotonin).

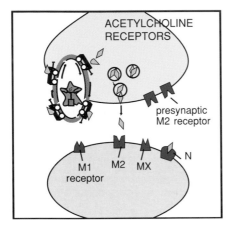

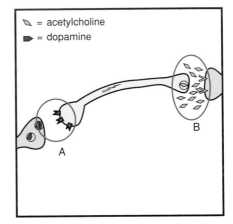

FIGURE 2.13. Acetylcholine (ACh) receptors. There are numerous receptors for ACh. The major subdivision is between nicotinic (N) and muscarinic (M) cholinergic receptors. There are also numerous subtypes of these receptors, best characterized for muscarinic receptor subtypes (M_1, M_2, M_X). Perhaps the M_1 postsynaptic receptor is key to mediating the memory functions linked to cholinergic neurotransmission, as well as the peripheral side effects of anticholinergic drugs, such as dry mouth, blurred vision, constipation and urinary retention.

FIGURE 2.14. Dopamine and acetylcholine have a reciprocal relationship in the nigrostriatal dopamine pathway. Dopamine neurons here make postsynaptic connections with cholinergic neurons. In the absence of dopamine (A), the cholinergic neuron is active (B).

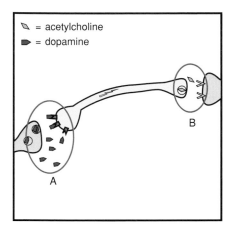

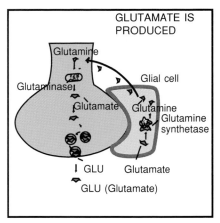

FIGURE 2.15. This figure shows what happens to acetylcholine activity when dopamine receptors are stimulated. Dopamine normally suppresses acetylcholine activity. Thus here, in the presence of dopamine (A), cholinergic output is reduced (B).

FIGURE 2.16. Glutamate is produced (synthesized). Glutamate or glutamic acid (Glu) is an amino acid that is a neurotransmitter. Its predominant use is not as a neurotransmitter, but as a building block (amino acid) of protein synthesis. When used as a neurotransmitter, it is synthesized from glutamine. Glutamine is converted into glutamate by an enzyme in mitochondria called glutaminase. It is then stored in synaptic vesicles for subsequent release during neurotransmission. Glutamine itself can be obtained from glia cells adjacent to neurons. Glia cells have a supportive role for neurons, supporting them both structurally and metabolically. In the case of glutamate neurons, nearby glia can provide glutamine for neurotransmitter glutamate synthesis. In this case, glutamate from metabolic pools in the glia is converted into glutamate use as a neurotransmitter. This is accomplished by first converting glutamate into glutamine in the glia cell via the enzyme glutamine synthetase. Glutamine is then transported into the neuron for conversion into glutamate and use as a neurotransmitter.

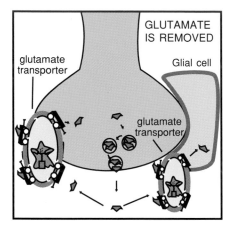

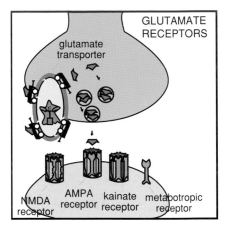

FIGURE 2.17.   Glutamate removal. Glutamate's actions are stopped not by enzymatic breakdown, as in other neurotransmitter systems, but by removal by two transport pumps (glutamate transporter). The first of these pumps is a presynaptic glutamate transporter which works like all the other neurotransmitter transporters already discussed for amine neurotransmitter systems, such as dopamine, serotonin and choline. The second transport pump is located on nearby glia, which removes glutamate from the synapse and terminates its actions there.

FIGURE 2.18.   Glutamate receptors. There are several types of glutamate receptors, including three that are linked to ion channels: NMDA (*N*-methyl-D-aspartate), AMPA (α-amino-3-hydroxy-5-methyl-4-isoxazole-propionic acid), and kainate, all named after the agonists that selectively bind to them. Another type of glutamate receptor is the metabotropic glutamate receptor, which is a G-protein-linked receptor; this may mediate long-lasting electrical signals in the brain called 'long-term potentiation'. Long-term potentiation appears to have a key role in memory functions.

For more information on glutamate see: Bloom and Kupfer 1996; Cooper et al 1996; Stahl 1996.

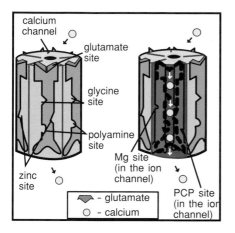

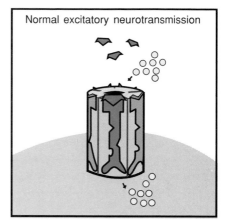

FIGURE 2.19. Five modulatory sites on the NMDA (N-methyl-D-aspartate) receptor. The NMDA glutamate-calcium channel complex has multiple receptors in and around it which act in concert as allosteric modulators. Three of these modulatory sites are located around the NMDA receptor and are shown on the left. One of these modulatory sites is for the neurotransmitter glycine, another is for polyamines, and yet another for zinc.

Two of the modulatory sites are located inside or near to the ion channel itself, shown on the right. The ion magnesium can block the calcium channel at one of these modulatory sites, presumably inside the ion channel or closely related to it. The other inhibitory modulatory site located inside the ion channel is sometimes called the 'PCP site' because the psychotomimetic agent PCP (i.e., phencyclidine) binds to this site.

FIGURE 2.20. Normal excitatory neurotransmission at the NMDA (N-methyl-D-aspartate) type of glutamate receptor. The NMDA receptor is a ligand-gated ion channel. This fast-transmitting ion channel is an excitatory calcium channel. Occupancy of NMDA glutamate receptors by glutamate causes calcium channels to open and the neuron to be excited for neurotransmission.

CHAPTER 3

SYMPTOMS AND PATHOPHYSIOLOGY OF SCHIZOPHRENIA

Discovery of the antipsychotics in the 1950s led to an emphasis on the positive symptoms of schizophrenia, such as delusions and hallucinations, which the conventional antipsychotics so dramatically reduced. However, reduction of these symptoms did not lead to recovery from schizophrenia.

Other symptoms of schizophrenia include so-called negative symptoms, such as apathy, withdrawal and lack of pleasure. The new atypical antipsychotics reduce such negative symptoms to a greater extent than do the conventional antipsychotics. However, all symptom dimensions of schizophrenia are not adequately described by the combination of positive and negative symptoms. Some research now classifies schizophrenia into five symptom dimensions: positive and negative as well as cognitive, aggressive/hostile and depressive/anxious.

This chapter describes these five domains of symptoms in schizophrenia as well as leading hypotheses describing the pathophysiology of the different symptoms and the possible etiology of schizophrenia.

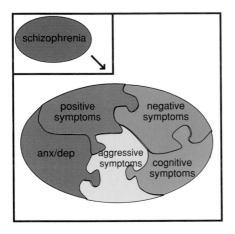

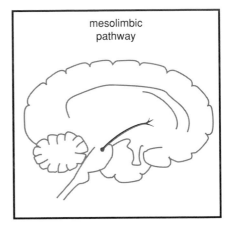

FIGURE 3.1. Schizophrenia has been classified predominantly as a disorder of positive and negative symptoms. These will be described in the following figures and tables. Note that not all symptoms of schizophrenia can be described by positive and negative symptoms. These other symptoms are discussed after a description and explanation of positive and negative symptoms.

For further reading see: Lieberman et al 1997; Marder et al 1997; Schooler 1994.

FIGURE 3.2. The mesolimbic dopamine pathway, projecting from the ventral tegmental area to limbic regions, including the nucleus accumbens, is shown here. An enlarged version of the terminal projection, with functional activity, is shown in Figure 3.3.

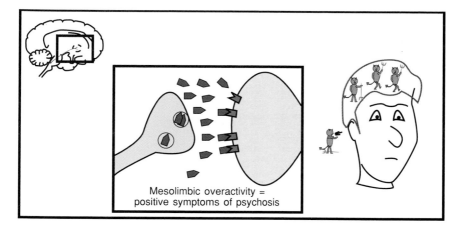

FIGURE 3.3. The dopamine hypothesis of psychosis. Overactivity of dopamine neurons in the mesolimbic dopamine pathway may mediate the positive symptoms of psychosis.

For further reading see: Cooper et al 1996; Kahn and Davis 1996; Knable et al 1998; Meltzer and Stahl 1976; Stahl 1996.

Table 3.1
Positive symptoms of psychosis

Delusions
Hallucinations
Distortions or exaggerations in
    language and communication
Disorganized speech
Disorganized behavior
Catatonic behavior
Agitation

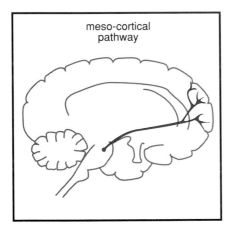

meso-cortical pathway

FIGURE 3.4. The mesocortical pathway, like the mesolimbic pathway shown in Figure 3.2, originates in the ventral tegmental area of the brain stem. It projects to limbic cortex. An enlarged version of the terminal projection, with functional activity, is shown in Figure 3.5.

Table 3.2
Negative symptoms of psychosis

Blunted affect
Emotional withdrawal
Poor rapport
Passivity
Apathetic social withdrawal
Difficulty in abstract thinking
Lack of spontaneity
Stereotyped thinking
Alogia: restrictions in the fluency and productivity of thought and speech
Avolition: restrictions in the initiation of goal-directed behavior
Anhedonia: lack of pleasure
Attentional impairment

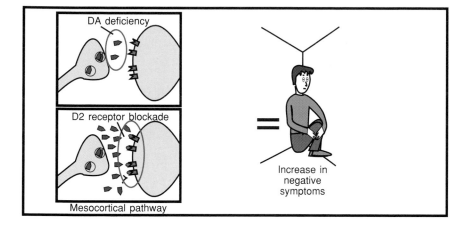

FIGURE 3.5. The mesocortical pathway may mediate certain cognitive symptoms, especially the negative symptoms of schizophrenia.

Table 3.3 Negative symptoms etiology
Cortical dopamine deficiency? Caused by mesocortical dopamine blockade? Worsened by mesocortical dopamine blockade?

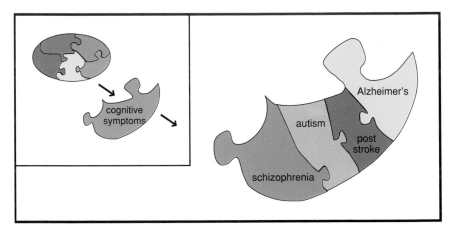

FIGURE 3.6. In addition to the positive and negative symptoms of schizophrenia, cognitive symptoms are associated with schizophrenia as well as other cognitive disorders. Cognitive symptoms associated with schizophrenia are listed in Table 3.4.

For further reading see: Goldberg and Weinberger 1996; Harvey and Keefe 1997.

Table 3.4 Cognitive problems
Thought disorder Incoherence Loose associations Neologisms Impaired attention Impaired information processing

Table 3.5 Several of the most severe cognitive impairments in schizophrenia
Impaired verbal fluency (ability to produce spontaneous speech) Serial learning (of a list of items or a sequence of events) Impairment in vigilance for executive functioning (problems with sustaining and focusing attention, concentrating, prioritizing and modulating behavior based on social cues)

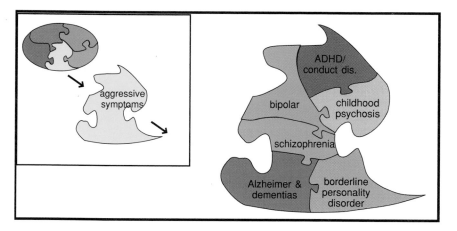

FIGURE 3.7. Another dimension of symptoms in schizophrenia is the aggressive symptoms, such as hostility, acting out towards self, others and property, and poor impulse control. These are associated with other disorders as well as schizophrenia.

Table 3.6
Aggressive symptoms

Hostility
Verbal abusiveness
Physical abusiveness
Assault
Self-injurious behavior including
    suicide
Arson/property damage
Impulsiveness
Sexual acting out

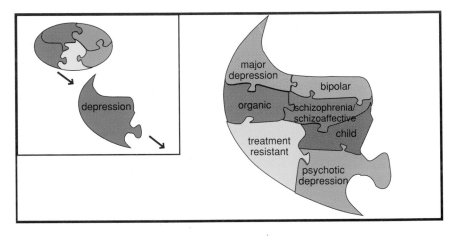

FIGURE 3.8. A fifth dimension of the symptoms in schizophrenia includes symptoms of depression and anxiety. This does not necessarily mean a co-morbid affective or anxiety disorder fulfilling full DSM-IV (the *Diagnostic and Statistical Manual of Mental Disorders*, 4th edition) diagnostic criteria, but nevertheless there are symptoms of depressed and anxious mood associated with schizophrenia as well as other disorders.

Table 3.7
Depressive/anxious symptoms

Depressed mood
Anxious mood
Guilt
Tension
Irritability
Worry

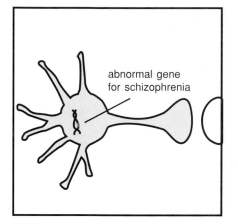

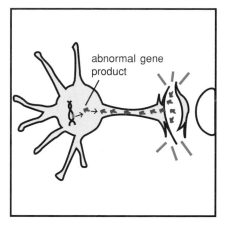

FIGURE 3.9. This figure shows the postulated abnormal gene for schizophrenia lying latent in the cell. In this case it does not produce abnormal gene products or cause schizophrenia.

For further reading see: Knable et al 1998; Stahl 1996.

FIGURE 3.10. Here the postulated abnormal gene for schizophrenia is being expressed, leading to an abnormal gene product which causes disruption in the functioning of the neuron; this, in turn, leads to psychosis and the other symptoms of schizophrenia.

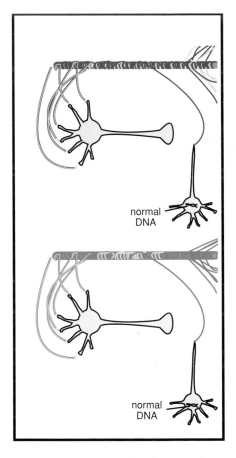

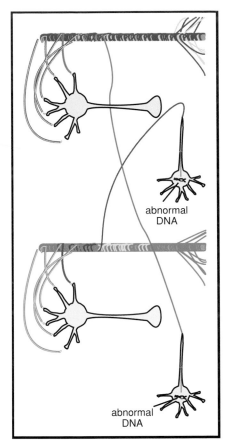

FIGURE 3.11. Neurodevelopmental theories of schizophrenia suggest that something goes wrong with the program for the normal formation of synapses and migration of neurons in the formation of the brain and its connections during their prenatal and early childhood stages. This figure depicts a concept of how a neuron with normal DNA would develop and form synaptic connections.

FIGURE 3.12. According to neurodevelopmental theories of schizophrenia, an abnormality in the DNA of a schizophrenic patient may cause the wrong synaptic connections to be made in the formation of the brain and its connections during their prenatal and early childhood stages. Schizophrenia may be the result of abnormal development of the brain from the beginning of life, because neurons fail to migrate to the correct parts of the brain, fail to form appropriate connections, and then break down when used by the individual in late adolescence and early adulthood.

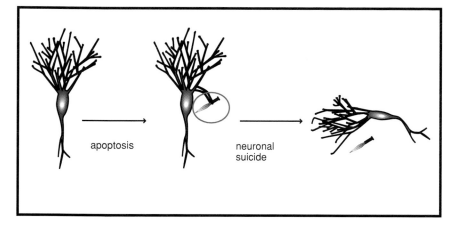

apoptosis

neuronal
suicide

FIGURE 3.13. Apoptosis. Neurons are equipped with the molecular machinery to commit suicide by a process called apoptosis (cell death). Although necrosis (see Figure 3.15) is a messy affair characterized by a severe and sudden injury with an inflammatory response, apoptosis is more subtle and akin to fading away. Apoptosis literally means 'falling off' as the petals fall off a flower or the leaves fall from a tree. Apoptosis is a natural part of development of the immature central nervous system. The brain originally makes excess neurons and, during development, the fittest are selected and the rest fade away through apoptosis. Apoptosis is thus a natural mechanism to eliminate the unwanted neurons without the molecular mess of necrosis.

If the wrong cells are eliminated by faulty programming, or by *in utero* chemicals and toxins, or even by undesirable experiences, aberrant neurodevelopment may occur. This is one hypothesis for the pathophysiology of schizophrenia, namely that inappropriate apoptosis during neurodevelopment selects the wrong neurons, with the consequence that the wrong connections are made between neurons; ultimately, this leads to the neurodevelopmental disorder of schizophrenia.

For further reading see: Lewis 1997; Liebermann et al 1997.

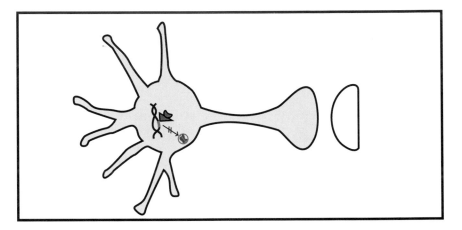

FIGURE 3.14. Genetic approach to therapeutics in schizophrenia based upon molecular neurobiology. If a degenerative process is 'turned on' genetically at the beginning of the course of schizophrenia (see Figure 3.10), perhaps it could be 'turned off' pharmacologically by a drug that is able to prevent the expression of the postulated abnormal gene product in schizophrenia. This could theoretically arrest the disease and prevent it from further progression.

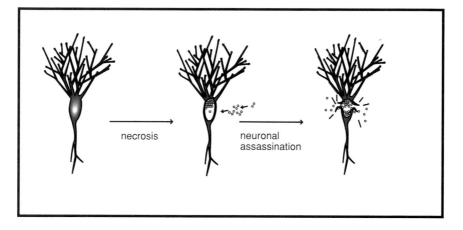

FIGURE 3.15. Necrosis. If apoptosis is fading away, necrosis is a dramatic battle. Apoptotic cells shrink, whereas necrotic cells explode. If apoptosis is cellular suicide, necrosis is cellular assassination. Three common neuronal assailants are poisons, ischemia and abusive overwork. Perhaps glutamate can trigger necrosis (and/or apoptosis) by an excitotoxic mechanism which is explained in Figures 3.16-3.20.

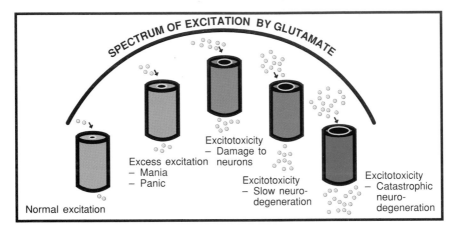

FIGURE 3.16. Neuroprotection, excitotoxicity and the glutamate system in degenerative disorders. A major research strategy for the discovery of novel therapies for neurodegenerative disorders is to target the glutamate system, which might mediate progressive neurodegeneration by an excitotoxic mechanism. Such an excitotoxic mechanism may play a role in various neurodegenerative diseases such as schizophrenia, Alzheimer's disease, Parkinson's disease, Huntington's disease, amyotrophic lateral sclerosis and even stroke. The spectrum of excitation by glutamate ranges from normal neurotransmission, to excess neurotransmission causing pathologic symptoms such as mania or panic, to excitotoxicity resulting in minor damage to dendrites, to slow progressive excitotoxicity resulting in neuronal degeneration such as in Alzheimer's disease, to sudden and catastrophic excitotoxicity causing neurodegeneration as in stroke.

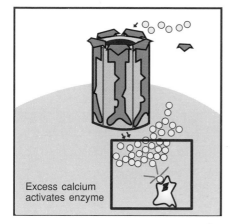

FIGURE 3.17. Cellular events occurring during excitotoxicity (part 1). Excitotoxicity is a major current hypothesis for explaining a neuropathologic mechanism that could mediate the final common pathway of any number of neurologic and psychiatric disorders characterized by a neurodegenerative course. The basic idea is that the normal process of excitatory neurotransmission runs amok and, instead of normal excitatory neurotransmission, things get out of hand, and the neuron is literally excited to death.

The excitotoxic mechanism is thought to begin with a pathologic process which triggers excessive glutamate activity. This causes excessive opening of the calcium channel, shown here, with poisoning of the cell by this excessive calcium.

FIGURE 3.18. Cellular events occurring during excitotoxicity (part 2). Once excessive glutamate causes too much calcium to enter the neuron, the next stage is for the calcium to activate enzymes that produce free radicals. Free radicals are chemicals that are capable of destroying other chemicals and cellular components.

FIGURE 3.19. Cellular events occurring during excitotoxicity (part 3). As the calcium accumulates in the cell, and the enzymes produce more and more free radicals, they begin to destroy parts of the cell indiscriminately, especially its membrane and organelles such as energy-producing mitochondria.

FIGURE 3.20. Cellular events occurring during excitotoxicity (part 4). Eventually, the damage is so great that the free radicals essentially destroy the whole neuron.

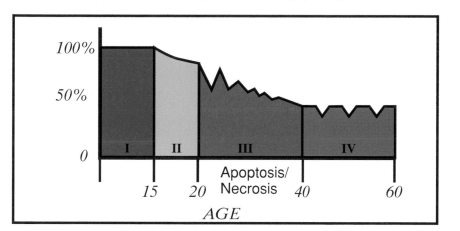

FIGURE 3.21. Schizophrenia as a neurodegenerative disorder. The natural history of schizophrenia, depicting the classic downhill course, is shown here in several stages: the first is the premorbid stage, in which functioning is normal (I). Next is a prodromal state characterized by subtle symptoms but little or no loss of functioning (II). Low levels of negative symptoms can have their onset in this stage. The prodromal stage then leads to chaotic symptoms of mental illness, which suddenly disrupt social and occupational function (III). They also hypothetically trigger apoptosis or necrosis, which is destructive both to the brain and to the level of function in an ultimately progressive process. Finally, a 'burn-out' stage may emerge in which chaotic symptoms of mental illness are gone, but residual negative and cognitive symptoms, as well as treatment nonresponsiveness, predominate (IV).

For further reading see: Lieberman et al 1997.

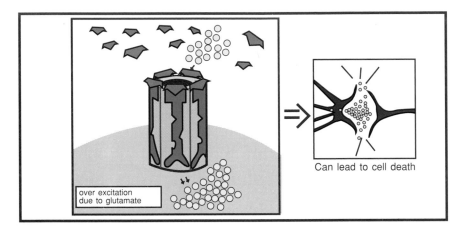

FIGURE 3.22. Excessive glutamate
activity is shown here to be mediated by
excesses in the amounts of glutamate
itself. Left to proceed too far, this could
lead to cell death from excitotoxicity.

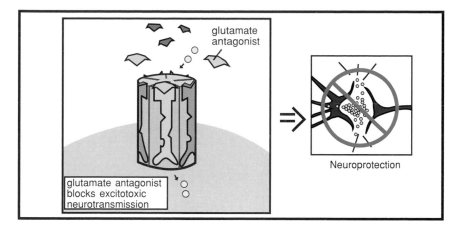

FIGURE 3.23. Antagonists of glutamate
at the agonist site can block excitotoxic
neurotransmission, and exert
neuroprotective actions.

FIGURE 3.24. Free radicals are generated in the neurodegenerative process of excitotoxicity. A drug acting as a free radical 'scavenger', soaking up toxic free radicals like a chemical sponge and removing them, would be neuroprotective. Vitamin E is a weak scavenger. Other free radical scavengers, such as the 'lazaroids' (so named because of their putative Lazarus-like actions of raising degenerating neurons from the dead), are also being tested.

CHAPTER 4

CONVENTIONAL ANTIPSYCHOTICS: THE CLASSICAL NEUROLEPTICS

The first antipsychotic drugs were discovered by accident in the 1950s when a drug that was thought to be an antihistamine (chlorpromazine) was serendipitously observed to have unique antipsychotic effects, when the putative 'antihistamine' was tested in schizophrenia patients. Chlorpromazine indeed has some antihistaminic activity, but also has more important activity at dopamine receptors. It even has additional – but generally unwanted – activity at α_1-adrenergic receptors and muscarinic/cholinergic receptors.

Once chlorpromazine was observed to be an effective antipsychotic agent, it was tested experimentally to try to discover its mechanism of action. Early in the testing process, chlorpromazine and other antipsychotic agents were all found to cause 'neurolepsis' in experimental animals. Thus, the conventional antipsychotics are sometimes referred to as 'neuroleptics'. New antipsychotics were discovered largely by their ability to produce this effect. It was not until many years later, perhaps in the late 1960s and 1970s, that it was widely recognized that all the known antipsychotics at that time shared the common property of blocking dopamine-2 receptors.

The various conventional antipsychotics differ in terms of their ability to block histamine, α_1- and muscarinic/cholinergic receptors, but not so much in terms of their ability to block dopamine-2 receptors. Notably, the conventional antipsychotics are largely devoid of blockade of serotonin-2_A ($5HT_{2A}$) receptors, a key property of the new atypical antipsychotics which will be discussed in Chapter 5. As a result of binding properties of the conventional antipsychotics, they differ in their side-effect profiles, but not overall in their therapeutic profiles. Some conventional antipsychotics are more sedating than others, some have more ability to cause cardiovascular side effects than others, and some are more potent than others. However, all reduce psychotic symptoms – especially positive psychotic symptoms – about equally in groups of schizophrenic patients in multicenter trials. This is not to say that one individual patient might not respond better to one conventional antipsychotic agent rather than another, but there is no recognized difference in the efficacies among all the conventional antipsychotic agents when tested in large groups of patients. This is not the case for the atypical antipsychotics which are recognized as having superior efficacy, particularly for negative symptoms, when compared with conventional antipsychotics. The atypical antipsychotic clozapine has the best-documented efficacy for treating schizophrenia when two or more conventional antipsychotics fail to give an adequate result, that is, in the treatment-refractory group of patients. This will be covered in Chapter 6.

Here, we discuss the conventional antipsychotic agents, and show how they are all capable of reducing positive symptoms of psychosis, as well as producing undesirable extrapyramidal symptoms (EPS) and tardive dyskinesia. All these properties derive from the dopamine-2 (D_2) receptor-blocking properties of the conventional antipsychotics. Thus, this D_2-receptor antagonism mediates not only the therapeutic effects of conventional antipsychotic agents, but also some of the side effects of these very same agents. This chapter sets the stage for comparisons with the atypical antipsychotics, from which conventional antipsychotics differ in that the conventional antipsychotics are inferior to the atypical antipsychotics in both tolerability and efficacy.

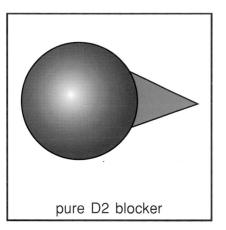

FIGURE 4.1. This figure represents a conventional antipsychotic drug. Such drugs have at least four actions: blockade of dopamine-2 receptors (D_2); blockade of muscarinic/cholinergic receptors (M_1); blockade of α_1-adrenergic receptors (α_1); and blockade of histamine receptors (antihistaminic actions; H_1).

For further reading: Baldessarini 1996; Bond and Lader 1996; Cooper et al 1996; Gelenberg and Bassuk 1997; Hyman et al 1995; Janicak et al 1997; Leonard 1997; Quitken et al 1998; Schatzberg and Nemeroff 1998; Stahl 1996.

FIGURE 4.2. Pure dopamine-2 (D_2) blocker. This icon represents the notion of a single pharmacologic action, namely D_2-receptor antagonism. Although actual drugs have multiple pharmacologic actions, this single action idea will be applied conceptually in several of the following figures.

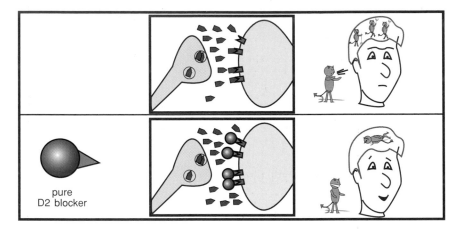

FIGURE 4.3. The dopamine receptor antagonist hypothesis of antipsychotic drug action. Blockade of postsynaptic dopamine receptors in the mesolimbic pathway is thought to mediate the antipsychotic efficacy of the antipsychotic drugs, and their ability to diminish or block positive psychotic symptoms.

For further reading see Meltzer and Stahl 1976; Stahl 1996.

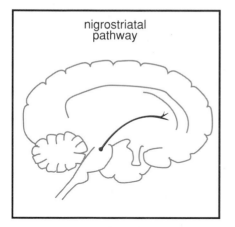

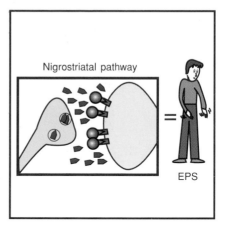

FIGURE 4.4. The nigrostrial dopamine pathway projects from the substantia nigra to the basal ganglia, and is thought to control movements.

FIGURE 4.5. When dopamine receptors are blocked in the postsynaptic projections of the nigrostriatal pathway, it produces disorders of movement which can appear very much like those in Parkinson's disease; this is why these movements are sometimes called drug-induced parkinsonism. As the nigrostriatal pathway projects to basal ganglia, a part of the extrapyramidal neuronal system of the CNS, side effects associated with blockade of dopamine receptors are sometimes also called extrapyramidal symptoms, or EPS.

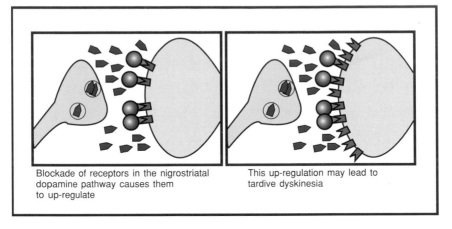

Blockade of receptors in the nigrostriatal dopamine pathway causes them to up-regulate

This up-regulation may lead to tardive dyskinesia

FIGURE 4.6. Long-term blockade of dopamine receptors in the nigrostriatal dopamine pathway may cause them to upregulate. A clinical consequence of this may be the hyperkinetic movement disorder known as tardive dyskinesia. This upregulation may be the consequence of the neuron's futile attempt to overcome drug-induced blockade of its dopamine receptors.

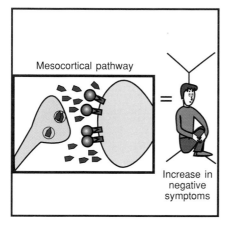

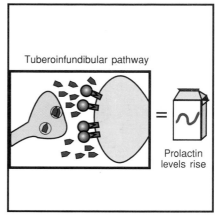

FIGURE 4.7. A pathway related to the mesolimbic dopamine pathway is the mesocortical dopamine pathway. It projects from the midbrain ventral tegmental area, but sends its axons to the limbic cortex, where it may have a role in mediating positive and negative psychotic symptoms or cognitive side effects of neuroleptic antipsychotic medications. When dopamine receptors are blocked in the mesocortical dopamine pathway, it may produce blunting of emotions and various cognitive side effects which actually mimic negative symptoms themselves. Sometimes these cognitive side effects of neuroleptics are called the 'neuroleptic-induced deficit syndrome'.

FIGURE 4.8. The fourth dopamine pathway of interest is the one that controls prolactin secretion, called the tuberoinfundibular dopamine pathway. It projects from the hypothalamus to the anterior pituitary gland. When the dopamine receptors in the tuberoinfundibular dopamine pathway are blocked, prolactin levels rise, sometimes so much so that women can begin lactating inappropriately, a condition known as galactorrhea.

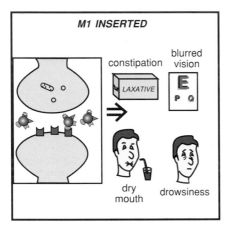

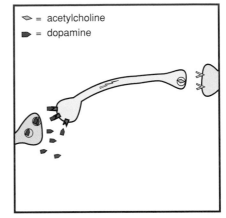

FIGURE 4.9. Side effects of the conventional antipsychotics (part 1). In this diagram, the icon of the conventional antipsychotic is shown with its M_1 (anticholinergic/antimuscarinic) portion inserted into acetylcholine receptors, causing the side effects of constipation, blurred vision, dry mouth and drowsiness.

FIGURE 4.10. Dopamine and acetylcholine have a reciprocal relationship in the nigrostriatal dopamine pathway. Dopamine neurons here make postsynaptic connections with cholinergic neurons. Normally, dopamine suppresses acetylcholine activity.

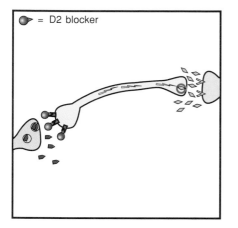

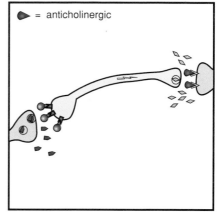

FIGURE 4.11. This figure shows what happens to acetylcholine activity when dopamine receptors are blocked. As dopamine normally suppresses acetylcholine activity, removal of this causes an increase in acetylcholine activity. Thus, if dopamine receptors are blocked, acetylcholine becomes overly active. This is associated with the production of extrapyramidal symptoms (EPS). The pharmacological mechanism of EPS therefore seems to be dopamine deficiency and acetylcholine excess.

FIGURE 4.12. One compensation for the overactivity of acetylcholine that occurs when dopamine receptors are blocked is to block the acetylcholine receptors with an anticholinergic agent. Thus, anticholinergics overcome excess acetylcholine activity caused by removal of dopamine inhibition when dopamine receptors are blocked by conventional antipsychotics. This also means that extrapyramidal symptoms (EPS) are reduced.

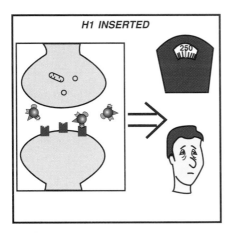

FIGURE 4.13. Side effects of the conventional antipsychotics (part 2). In this diagram, the icon of the conventional antipsychotic is shown with its H1 (antihistamine) portion inserted into histamine receptors, causing the side effects of weight gain and drowsiness.

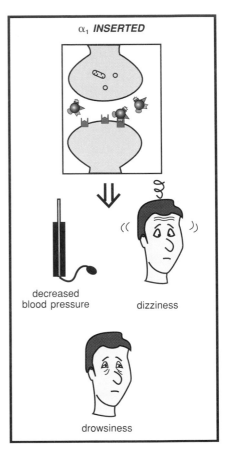

FIGURE 4.14. Side effects of the conventional antipsychotics (part 3). In this diagram, the icon of the conventional antipsychotic is shown with its α_1 (α_1-adrenergic antagonist) portion inserted into α_1-adrenergic receptors, causing the side effects of dizziness, decreased blood pressure and drowsiness.

Table 4.1
Conventional antipsychotic agents used to treat psychosis and schizophrenia in the USA

Generic name	Trade name	Action
Chlorpromazine	Thorazine	Typical neuroleptic
Fluphenazine	Prolixin; Permitil	Typical neuroleptic
Haloperidol	Haldol	Typical neuroleptic
Loxapine	Loxitane	Typical neuroleptic
Mesoridazine	Serentil	Typical neuroleptic
Molindone	Moban; Lidone	Typical neuroleptic
Perphenazine	Trilafon	Typical neuroleptic
Pimozide	Orap	Typical neuroleptic[a]
Prochlorperazine	Compazine	Typical neuroleptic[b]
Thioridazine	Mellaril	Typical neuroleptic
Thiothixene	Navane	Typical neuroleptic
Trifluperazine	Stelazine	Typical neuroleptic

[a]Approved in the USA for Gilles de la Tourette syndrome.
[b]Approved in the USA for nausea and vomiting, as well as psychosis.

Table 4.2
Clinical pearls about conventional antipsychotics

Inexpensive

May be as effective for positive symptoms as some atypical antipsychotics

Concomitant anticholinergics reduce EPS

Several agents are formulated for intramuscular administration for acute or depot use

Table 4.3
Rational explanation for conventional antipsychotic therapeutic effects

Blockade of dopamine-2 receptors in the mesolimbic pathway

Table 4.4
Rational explanation for conventional antipsychotic side effects: unwanted pharmacologic actions at secondary receptors

Blockade of dopamine-2 (D_2) receptors in the nigrostriatal pathway (EPS and tardive dyskinesia)

Blockade of D_2-receptors in the pathway (hyperprolactinemia with resultant galactorrhea, sexual dysfunction, gynecomastia, infertility, amenorrhea and possibly accelerated osteoporosis)

Blockade of D_2-receptors in the mesocortical pathway (production or worsening of negative symptoms)

Blockade of α_1-receptors (orthostatic hypotension, dizziness, sedation)

Blockade of muscarinic/cholinergic receptors (sedation, memory problems, dry mouth, blurred vision, constipation, urinary retention)

Blockade of histaminic receptors (weight gain and sedation)

Table 4.5
Preferred uses of
conventional antipsychotics

For patients stabilized long term
  with acceptable side effects
Second line when atypical
  antipsychotics fail
For 'top up' of patients receiving
  maintenance atypical
  antipsychotics who require
  intermittent treatment of
  aggression, as required

Table 4.7
Least preferred uses for
conventional antipsychotics

First-line treatment
Patients with tardive dyskinesia
Nonschizophrenic conditions
Children
Elderly people
Long-term maintenance

Table 4.6
Side-effect profile of
conventional antipsychotics

Parkinsonism
Extrapyramidal symptoms
Tardive dyskinesia
Memory problems
Neuroleptic-induced deficit
  syndrome
Dry mouth, blurred vision,
  constipation
Weight gain
Sedation
Orthostatic hypotension
Galactorrhea/amenorrhea
Sexual dysfunction

CHAPTER 5

ATYPICAL ANTIPSYCHOTICS AND SEROTONIN-DOPAMINE ANTAGONISM

This chapter covers the atypical antipsychotics, sometimes also called serotonin-dopamine antagonists (SDAs). The term 'atypical antipsychotic' has caused a good deal of debate and confusion, because it can mean different things to different experts: to prescribers it can connote 'low extrapyramidal symptoms' or 'good for negative symptoms'; to a pharmacologist, 'serotonin 2_A dopamine 2 antagonism'; to a marketeer, 'new and different'; to a formulary committee or payer with a short-term perspective, 'expensive', but, to a pharmacoeconomist with a long-term perspective, 'cost-effective'. Each of these dimensions of the atypical antipsychotics is addressed in this chapter.

The atypical antipsychotics will be presented as a group in this chapter in order to contrast them with the group of conventional antipsychotics presented in Chapter 4. There is general agreement that three features differentiate the atypical antipsychotics from the conventional antipsychotics, and that these features may arise from the pharmacologic property of serotonin-2_A/dopamine-2 ($5HT_{2A}/D_2$) antagonism which all atypical antipsychotics share, as opposed to D_2 antagonism without $5HT_{2A}$ antagonism, which almost all the conventional antipsychotics share. The first of these differentiating features is that the atypical antipyschotics show little or no propensity to cause extrapyramidal symptoms (EPS) or tardive dyskinesia, which are the most troublesome side effects of the conventional antipsychotics. Second, some atypical antipyschotics do not raise prolactin levels, which every conventional antipsychotic does. Finally, most atypical antipsychotics reduce negative symptoms of schizophrenia to a greater extent than the conventional antipsychotics.

After explaining the pharmacologic and clinical significance of atypical antipsychotics as a group, we then profile the differentiating features of each individual member of the atypical antipsychotic class in Chapter 6. This includes clozapine, risperidone, olanzapine and quetiapine which are all marketed world wide, as well as sertindole, which is not available in the USA and is 'on hold' from marketing elsewhere, and ziprasidone, which is in late clinical testing as of this writing.

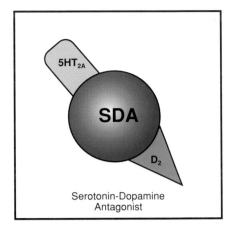

FIGURE 5.1. Serotonin-dopamine antagonist icon. This icon shows the dual pharmacologic actions which define serotonin-dopamine antagonists (SDAs), namely blockade of serotonin-2$_A$ (5HT$_{2A}$) receptors as well as dopamine-2 receptors.

For further reading see: Arnt and Skarsfeldt 1998; Kapur 1996; Kapur and Remington 1996; Kinon and Lieberman 1996; Mansour et al 1998; Meltzer 1996; Meltzer et al 1989; Owens and Risch 1998; Richelson 1996; Roth et al 1994; Stahl 1996; Stahl 1998a.

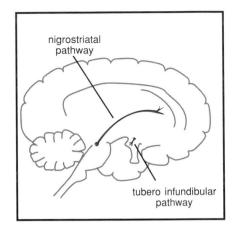

FIGURE 5.2. Reciprocal relationship between serotonin and dopamine in two dopamine pathways. Serotonin opposes the release of dopamine in two of the dopamine pathways shown here, namely the nigrostriatal pathway and the tuberoinfundibular pathway.

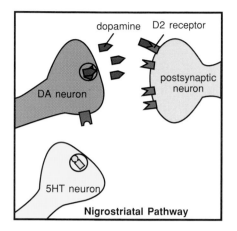

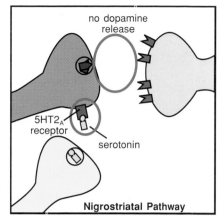

FIGURE 5.3. Enlarged view of serotonin (5HT) and dopamine (DA) interactions in the striatum. This pathway from substantia nigra to the striatum (basal ganglia) regulates movements. Dopamine release from this pathway is regulated by serotonin. The reciprocal relationship between dopamine and serotonin in the striatum was introduced in Chapter 2 (see Figure 2.10). Here, no serotonin (5HT) is present at its $5HT_{2A}$-receptor on the nigrostriatal dopaminergic neuron. Therefore, dopamine is being released.

FIGURE 5.4. Dopamine release is inhibited by serotonin (5HT). In contrast to Figure 5.3, dopamine release is being inhibited here. This lack of dopamine release is the result of serotonin's inhibitory actions when serotonin binds to $5HT_{2A}$-receptors on the nigrostriatal dopaminergic axon terminal.

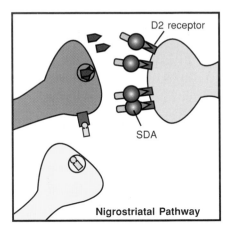

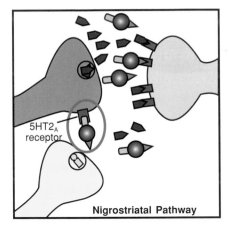

FIGURE 5.5. Dopamine receptors are blocked by serotonin-dopamine antagonists (SDAs). This figure shows what would happen if only the dopamine-2 (D$_2$) blocking action of a serotonin-dopamine antagonist were active: namely, the SDA molecule would only bind to postsynaptic D$_2$-receptors and block them.

FIGURE 5.6. Reversal of dopamine receptor blockade by serotonin-dopamine antagonists (SDAs). In contrast to Figure 5.5, this figure shows the dual action of the SDAs: the first action is to bind to the D$_2$-receptor as shown in Figure 5.5; the second is to bind to the 5HT$_{2A}$-receptor as shown here.

The interesting thing is that this second action actually reverses the first; that is, blocking a 5HT$_{2A}$-receptor reverses the blockade of a D$_2$-receptor. This happens because dopamine is released when serotonin can no longer inhibit this dopamine release. Another term for this is 'disinhibition'. Thus, blocking a 5HT$_{2A}$-receptor disinhibits the dopamine neuron, causing dopamine to pour out of the dopamine neuron. The consequence of this disinhibition is that the dopamine can then compete with the SDA for the D$_2$-receptor, and stop the inhibition there.

This is why 5HT$_{2A}$ blockers reverse D$_2$ blockers in the striatum. As D$_2$ blockade is thereby reversed, SDAs cause little or no extrapyramidal symptoms or tardive dyskinesia.

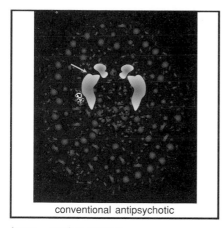

conventional antipsychotic

Arrow = caudate nucleus

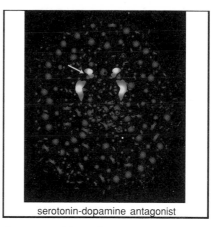

serotonin-dopamine antagonist

Arrow = caudate nucleus

FIGURE 5.7. Positron emission tomography (PET) scan of a conventional antipsychotic binding to dopamine receptors in the striatum. Bright colors indicate binding to D_2-receptors in this PET scan of a schizophrenic patient receiving a conventional antipsychotic at a clinically active dose. About 90% of dopamine receptors are being blocked, which explains why such doses cause extrapyramidal symptoms as they relieve psychosis.

FIGURE 5.8. PET scan of a serotonin-dopamine antagonist (SDA) binding to dopamine receptors in the striatum. Here there is less intense binding of the SDA to D_2-receptors in the striatum at a clinically active dose. Depending upon the dose of the SDA, less than 70 to 80% of D_2 receptors may be blocked. This explains why antipsychotic doses of an SDA may not be associated with extrapyramidal side effects. This scan proves in vivo that the $5HT_2$ antagonist action does reverse the D_2 antagonist action in the striatum, as previously shown in Figures 5.5 and 5.6.

For further reading see: Bench et al 1996; Kasper et al 1998; Kufferle et al 1997; Nyberg et al 1993; Nyberg et al 1997; Pilowsky et al 1996.

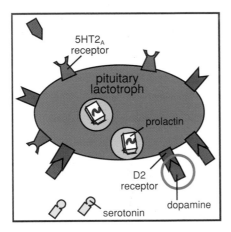

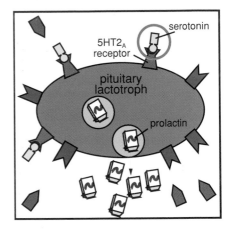

FIGURE 5.9. Enlarged view of dopamine and serotonin in the tuberoinfundibular pathway: dopamine inhibition of prolactin secretion. The dopamine pathway from hypothalamus to pituitary, called the tuberoinfundibular dopamine pathway, regulates prolactin release. Here, dopamine is occupying dopamine-2 (D_2) receptors, and prolactin release is *inhibited*.

FIGURE 5.10 Enlarged view of dopamine and serotonin in the tuberoinfundibular pathway: serotonin stimulation of prolactin secretion. In contrast to dopamine (see Figure 5.9), serotonin *stimulates* prolactin secretion when it occupies serotonin-2 ($5HT_2$) receptors. This may be both a direct effect on the pituitary lactotroph cell, and an indirect action on dopamine, by preventing dopamine release.

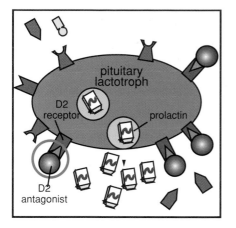

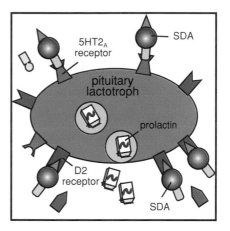

FIGURE 5.11. Effect of dopamine-2 (D$_2$) receptor blockade on prolactin secretion. As dopamine inhibits prolactin secretion, drugs that block D$_2$ receptors will *increase* prolactin levels.

FIGURE 5.12. How serotonin-2 (5HT$_{2A}$) antagonism reverses the ability of dopamine-2 (D$_2$) antagonism to increase prolactin secretion. As dopamine and serotonin have reciprocal regulatory roles in the control of prolactin secretion, one cancels the other. Thus, stimulating 5HT$_{2A}$-receptors reverses the effects of stimulating D$_2$-receptors (Figures 5.9 and 5.10). The same thing works in reverse, that is, blockade of 5HT$_{2A}$-receptors (shown here) *reverses* the effects of blocking D$_2$-receptors (compare Figure 5.11).

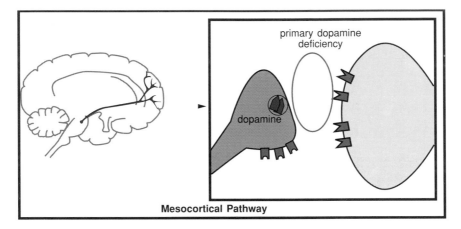

FIGURE 5.13. Primary deficiency of
dopamine in the mesocortical pathway,
a hypothetical cause of negative
symptoms of schizophrenia.

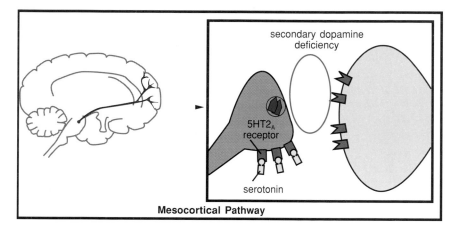

Mesocortical Pathway

FIGURE 5.14. Secondary deficiency of
dopamine in the mesocortical pathway
as a result of serotonin excess is another
hypothetical cause of negative symptoms
in schizophrenia.

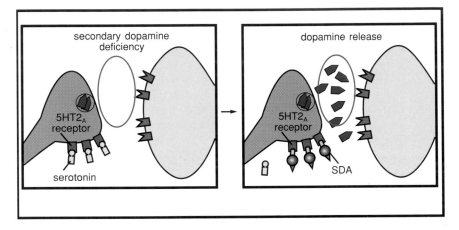

FIGURE 5.15. Serotonin-dopamine antagonists, but not conventional antipsychotics, can increase dopamine release selectively in the mesocortical pathway. This theoretically explains the improved efficacy of atypical antipsychotics over conventional antipsychotics in the treatment of negative symptoms of schizophrenia.

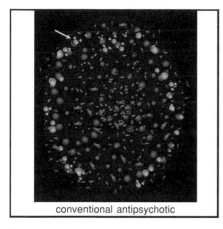

conventional antipsychotic

Arrow = cortex

FIGURE 5.16. PET scan of $5HT_{2A}$-receptor binding in a patient receiving a clinically effective dose of a conventional antipsychotic drug. As seen here, there is no binding of this kind of drug to $5HT_{2A}$-receptors.

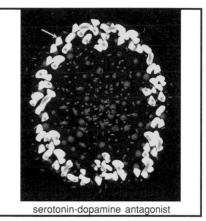

serotonin-dopamine antagonist

Arrow = cortex

FIGURE 5.17. PET scan of $5HT_{2A}$-receptor binding in a patient receiving a clinically effective dose of an atypical antipsychotic drug. In contrast to Figure 5.17, there is robust binding of these drugs to $5HT_{2A}$-receptors. Binding in key areas of limbic cortex may explain enhanced efficacy of atypical antipsychotic drugs in the treatment of negative symptoms of schizophrenia.

Table 5.1
New serotonin-dopamine antagonists

Risperidone
Olanzapine
Quetiapine

Table 5.2
Other serotonin-dopamine antagonists

Clozapine (the original)
Loxapine (one of the typical
    antipsychotics)
Zotepine (available in Europe)
Ziprasidone (in clinical
    development)
Sertindole ('on hold' from
    marketing)

Table 5.3
Rational explanations for SDA therapeutic effects

D_2-receptor blockade in the
    mesolimbic pathway to reduce
    positive symptoms
Enhanced dopamine release and
    $5HT_{2A}$-receptor blockade in the
    mesocortical pathway to reduce
    negative symptoms
Other receptor-binding properties
    may contribute to efficacy in
    treating cognitive symptoms,
    aggressive symptoms and
    depression in schizophrenia

Table 5.4
Rational explanation for SDA side-effect profile

$5HT_{2A}$ antagonism in the
    nigrostriatal pathway reduces
    EPS
$5HT_{2A}$ antagonism in the
    nigrostriatal pathway reduces
    tardive dyskinesia
$5HT_{2A}$ antagonism in the
    tuberoinfundibular pathway
    reduces hyperprolactinemia

**Table 5.5
Preferred uses for SDAs in
schizophrenia**

First-line treatment of positive
  symptoms
First-line treatment of negative
  symptoms
Treatment of relapses
For restabilization of patients
  experiencing side effects of
  conventional antipsychotics
Long-term maintenance to
  prevent relapse

**Table 5.6
Side-effect profile of SDAs**

Weight gain
Sedation
Insomnia
Agitation
Constipation
Dry mouth

**Table 5.7
Least preferred uses for
SDAs**

In combination with another
  atypical antipsychotic
At high doses (above
  recommended prescribing
  range is costly and reduces the
  atypical profile of some drugs)
For patients stabilized on
  conventional agents and who
  have an acceptable clinical
  response and side-effect profile
For parenteral (intramuscular)
  use (available only for loxapine;
  in development for ziprasidone,
  9-hydroxy-risperidone and
  olanzapine)

**Table 5.8
Clinical pearls about SDAs**

Can be expensive but cost-
  effective
More efficacy of many SDAs for
  negative symptoms compared
  to conventional antipsychotics
Not all SDAs have full atypical
  profile
Special uses in nonschizophrenic
  causes of positive psychosis
Special uses for cognitive and
  aggressive symptoms and
  depression in schizophrenia

**Table 5.9
Potential clinical benefits
deriving from the SDA
class**

Better compliance
Fewer hospitalizations
Less overall treatment costs
Less disruptive downhill course

**Table 5.10
Summary of the SDA
concept: how simultaneous
$5HT_{2A}/D_2$ antagonism may
differ from D_2 antagonism
alone**

Fewer EPS
Less tardive dyskinesia
Less prolactin elevation
Better negative symptom efficacy
Not all SDAs are atypical
SDAs which are atypical differ in
the degree to which the
atypical features apply (see
Chapter 6)

BEYOND THE SDA: HOW INDIVIDUAL ATYPICAL ANTIPSYCHOTICS DIFFER

Chapter 5 compared and contrasted serotonin – dopamine antagonists (SDAs) with the conventional antipsychotics presented in Chapter 4, through emphasis on generalizations about the SDAs as a class of drugs. Here, in Chapter 6, we describe how the concept of an SDA does not necessarily imply an 'atypical antipsychotic'. Furthermore, depending upon the definition chosen, individual drugs meet – to a greater or lesser extent – the different criteria for being an atypical antipsychotic. In this chapter, we attempt to answer whether eight different SDAs are atypical antipsychotics, namely: clozapine, risperidone, olanzapine, quetiapine, sertindole, ziprasidone, loxapine and zotepine. We also present the pharmacologic properties of these agents beyond antagonism of serotonin–dopamine, showing how no two antipsychotics have identical pharmacologic properties. Not surprisingly, each drug also has a correspondingly unique clinical profile, which will be described for all eight agents, based on differential pharmacology, results from clinical trials and real-world experience from clinical practice.

What is an atypical antipsychotic? Eight criteria are applied to this definition, in addition to the SDA properties:

1. Fewer EPS than with haloperidol
2. Essentially patients never develop EPS
3. Reduced incidence of tardive dyskinesia
4. Less prolactin than with haloperidol
5. Essentially no increases in prolactin
6. Better improvement of negative symptoms than with placebo
7. Better improvement of negative symptoms than with haloperidol
8. Effective for symptoms that are refractory to treatment with conventional antipsychotics.

Information about the drugs presented here is accruing at a fast and furious pace, requiring constant updating. Also, as clinical data emerge, contradictions can arise, compounded by marketing influences which tend to look at the same data from different perspectives. There are unfortunately few direct comparisons of atypical antipsychotics, and results from clinical trials do not necessarily agree with those from clinical practice. Here we have attempted to put together the consensus on these issues at the date of writing, based on the best available clinical studies that have been published and are soon to be published, together with anecdotal clinical observations. Thus, in addition to evaluating each of these eight agents for their atypical antipsychotic properties, we present 'clinical pearls' and dosage tips for the prescriber.

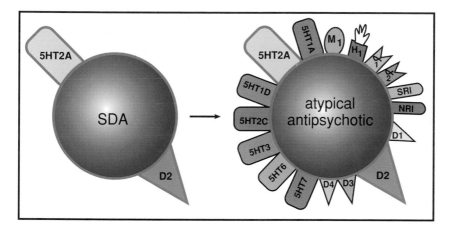

FIGURE 6.1. Beyond the SDA concept. Atypical antipsychotics are not simple serotonin–dopamine antagonists (SDAs). In truth, they have the most complex mixtures of pharmacologic properties in psychopharmacology. Shown here is an icon with all these properties. Beyond antagonism of serotonin-2 ($5HT_{2A}$)- and dopamine-2 (D_2)-receptors, there are a dozen or more other properties for drugs in the class of atypical antipsychotics. These agents act on multiple dopamine receptors (not just D_2, but also D_1, D_3 and D_4); on multiple serotonin receptors (not just $5HT_{2A}$, but also $5HT_{1A}$, $5HT_{1D}$, $5HT_{2C}$, $5HT_3$, $5HT_6$, $5HT_7$ and serotonin reuptake inhibition or SRI). Other neurotransmitter systems, which the atypical antipsychotics affect, include the noradrenergic system (α_1 and α_2 blockade), the cholinergic system (muscarinic blockade and norepinephrine reuptake inhibition or NRI) and antihistamine properties. Even more properties beyond those shown in this icon are being discovered at a rapid pace.

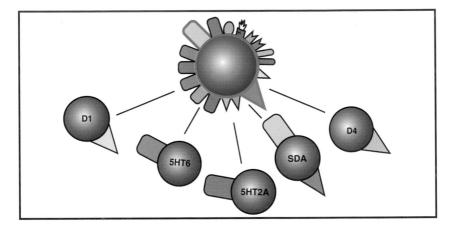

FIGURE 6.2. Which property is important? It became apparent during the 1980s that clozapine differs from the conventional antipsychotics in many ways. The pharmacologic mechanism of clozapine's special properties has been under intense investigation. As discussed in Chapter 5, clozapine's properties of essentially no EPS, tardive dyskinesia or prolactin elevations have been clearly linked in part to serotonin-dopamine antagonism. This is why many compounds that share this pharmacology also share these clinical characteristics of clozapine.

Less clear is why clozapine has the ability to treat patients refractory to numerous prior treatments with conventional antipsychotics. Enhanced efficacy of clozapine for negative symptoms may be related to serotonin-dopamine antagonism, but enhanced efficacy for treatment nonresponders, or for cognitive, aggressive and depressive symptoms, may be related to other pharmacologic properties.

Interestingly, attempts to find highly selective pharmacologic agents with improved clinical efficacy have been generally disappointing; that is, selective $5HT_{2A}$ antagonists and selective D_4 antagonists have not yet proved to be effective antipsychotic agents, although such agents are still being tested clinically. Thus, it seems that combinations of simultaneous actions may be required to explain the mechanism of enhanced efficacy of atypical antipsychotic drugs.

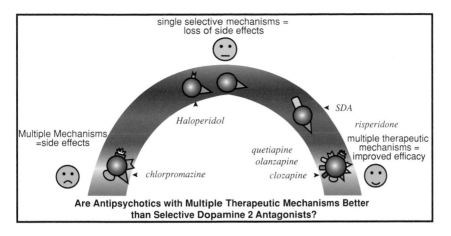

single selective mechanisms =
loss of side effects

Multiple Mechanisms
=side effects

Haloperidol

◀ SDA

risperidone

multiple therapeutic
mechanisms =
improved efficacy

quetiapine
olanzapine

clozapine ▶

◀ *chlorpromazine*

**Are Antipsychotics with Multiple Therapeutic Mechanisms Better
than Selective Dopamine 2 Antagonists?**

FIGURE 6.3. Are antipsychotics with multiple therapeutic mechanisms better than selective D_2 antagonists or selective SDAs? The original phenothiazines are conceptualized as conventional antipsychotics with the desirable pharmacologic property of D_2 antagonism, whereas its other pharmacologic properties are considered unwanted, and the cause of side effects (see left hand of the spectrum). Thus, when higher potency D_2 antagonists with lesser secondary pharmacologic properties were introduced, such as haloperidol, this was considered an advance (see middle of the spectrum). During this era, the idea was that the most desirable agents were those with the greatest selectivity and with only one primary action, namely D_2 antagonism.

All this has changed in the modern era of antipsychotics, where it is conceptualized that, at a minimum, $5HT_{2A}$ antagonism (SDA) should be combined with D_2 antagonism to make a more efficacious, better tolerated antipsychotic, namely an atypical antipsychotic. Taking things a step further is the proposition that even greater efficacy can be attained if a further mix of pharmacologic properties is present. Undoubtedly, some secondary pharmacologic properties are undesired, and account for side effects. On the other hand, it is clear that other combinations of pharmacologic mechanism may be synergistic, especially for efficacy in treatment-refractory schizophrenia, and for additional dimensions of symptoms in schizophrenia beyond positive and negative symptoms, such as cognitive, aggressive and depressive symptoms.

A new therapeutic goal of the emerging atypical antipsychotics is to mix a pharmacology of multiple neurotransmitter receptor actions that can reliably trigger 'awakenings' from schizophrenia and arrest the downhill course of illness (see Figure 6.1).

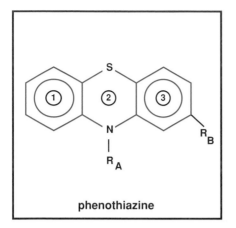

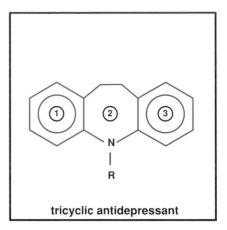

FIGURE 6.4. Structure of phenothiazines. The formula for the conventional antipsychotic phenothiazines is shown here in order to compare and contrast it with the formulae for other agents. Phenothiazines were first tested for their antihistamine properties. Only by serendipity were they discovered to have antipsychotic actions in the 1950s, and to work by blocking D_2-receptors in the 1970s. Many other chemical structures are included in the conventional antipsychotic class, and are given in the figures that follow.

FIGURE 6.5. Tricyclic structure. Early attempts to improve the antipsychotic actions of phenothiazines by modification of the chemical structure of the phenothiazines led to the synthesis of other structures with three rings. Although these modified tricyclic compounds lack antipsychotic effects, they have antidepressant actions, and are the chemical backbone of the tricyclic antidepressants.

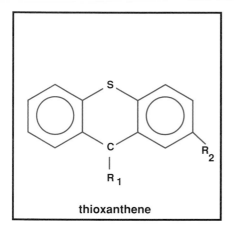

thioxanthene

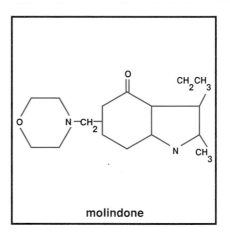

molindone

FIGURE 6.6. Structure of thioxanthines. This is another conventional antipsychotic chemical class that includes chlorprothixene and thiothixene.

FIGURE 6.7. Structure of another conventional antipsychotic: molindone.

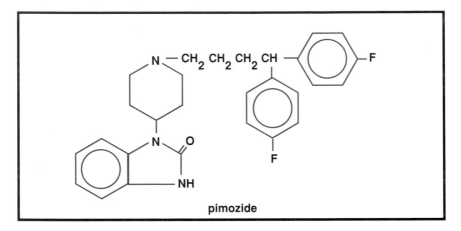

FIGURE 6.8. Structure of pimozide,
another conventional antipsychotic
agent.

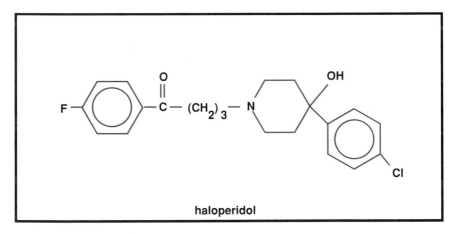

FIGURE 6.9. Structure of haloperidol,
one of the most widely prescribed
conventional antipsychotics during the
height of the conventional antipsychotic
era, before the mid 1990s.

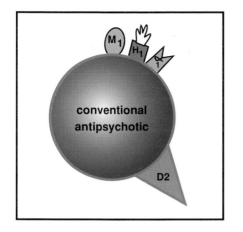

FIGURE 6.10.   Conventional antipsychotic pharmacologic icon. The most prominent binding properties of conventional antipsychotics of most chemical classes are represented (see also Figure 4.1).

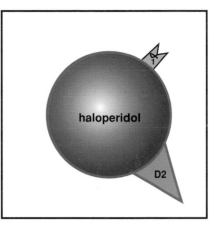

FIGURE 6.11.   Haloperidol pharmacologic icon. Haloperidol is different from many of the conventional antipsychotics in that it is more potent, and also lacks muscarinic and histaminic binding activities. Otherwise, its clinical profile is highly conventional.

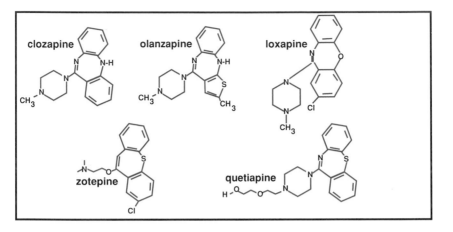

FIGURE 6.12.   Structure of clozapine and four other antipsychotics. Olanzapine, quetiapine, loxapine and zotepine have chemical structures related to clozapine; they are also SDAs. However, these five drugs have different clinical properties. Clozapine is in a class by itself (best efficacy, worst toxicity); olanzapine and quetiapine have well-documented atypical antipsychotic features, but loxapine and zotepine do not.

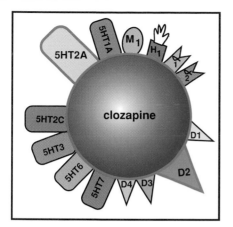

FIGURE 6.13. Clozapine pharmacologic icon. The most prominent binding properties of clozapine are represented here; it has perhaps the most complex binding portfolio for any drug in psychopharmacology. Binding properties vary greatly with technique and species, and from one laboratory to another. This icon portrays a qualitative consensus of current thinking about the binding properties of clozapine, which are constantly being revised and updated.

Table 6.1
Clozapine: The prototypical atypical

SDA?	Yes
Fewer EPS than with haloperidol?	Yes
Essentially patients never get EPS?	Yes[a]
Essentially patients never get tardive dyskinesia?	Yes
Less prolactin than with haloperidol?	Yes
Essentially prolactin never increases?	Yes
Better negative symptoms than with conventional antipsychotics?	Yes
Effective for symptoms that are refractory to conventional antipsychotics?	Yes

[a]Rare akathisia.

Table 6.2
Clinical pearls about clozapine

- Most efficacious but most dangerous
- May reduce violence and aggression in difficult cases
- May improve tardive dyskinesia
- Reduces suicide in schizophrenia
- Clinical improvements continue slowly over several years
- Not a first-line/first-break treatment choice in most countries
- Can cause agranulocytosis (0.5–2%)
- Monitoring of blood counts necessary weekly for 6 months, and then every 2 weeks
- Increased risk of seizures related to dose
- Doses over 550 mg/day may require concomitant anticonvulsant
- Can cause significant weight gain
- Sedation and sialorrhea (especially at night) may be bothersome

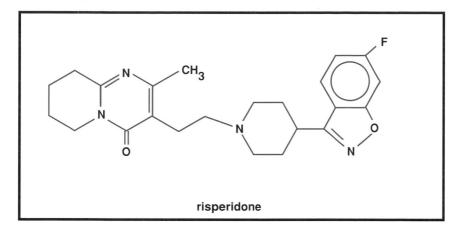

risperidone

FIGURE 6.14. Structure of risperidone. This was the first serotonin–dopamine antagonist to be marketed after the discovery of clozapine's atypical antipsychotic features.

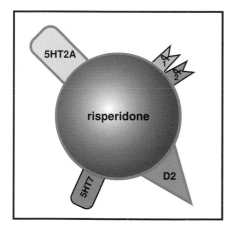

FIGURE 6.15. Risperidone pharmacologic icon, which portrays a qualitative consensus of the current thinking about the binding properties of risperidone. As for all atypical antipsychotics discussed in this chapter, binding properties vary greatly with technique and species, and from one laboratory to another; they are constantly being revised and updated.

Table 6.3 Risperidone: Is it atypical?	
SDA?	Yes
Fewer EPS than with haloperidol?	Yes[a]
Essentially patients never get EPS?	No[a]
Reduced incidence of tardive dyskinesia?	Probably
Less prolactin than with haloperidol?	No[b]
Better negative symptoms than with placebo?	Yes
Better negative symptoms than with 20 mg haloperidol?	Yes
Effective for symptoms refractory to conventional antipsychotics?	Maybe

[a]At low doses, EPS are the same as with placebo although they are occasionally seen; at high doses, EPS are increased, but still less than with haloperidol.
[b]Prolactin increases to the same extent or more than with haloperidol.

**Table 6.4
Clinical pearls about
risperidone**

- Well accepted for treatment of agitation and aggression in elderly demented patients

- Well accepted for treatment of bipolar disorders and schizophrenia

- Many anecdotal reports of use in children and treatment refractory cases, and for positive symptoms of psychosis in disorders other than schizophrenia

- Only atypical antipsychotic that elevates prolactin levels, although this is of unproven and uncertain clinical significance

- Although low doses cause no more EPS than placebo, this does not mean that they never cause EPS

- Less weight gain than with some other antipsychotics, although this does not mean that there is never weight gain

Table 6.5
Dosage tips for risperidone

- *Less may be more:* By lowering the dose, the side effects often reduced without loss of efficacy
- Thus, doses used in clinical practice are *lower* than doses suggested from early clinical trials
- Target dose for best efficacy/best tolerability may be 2–6 mg/day (average 4.5 mg/day) except in children or elderly people who should receive 0.5–2.0 mg/day
- Patients who respond to these doses may incur the lowest drug costs in the use of atypical antipsychotics
- Low doses may not be adequate in difficult patients
- Rather than raise the dose above these levels in agitated patients, partial responders or acutely ill patients requiring antipsychotic actions, consider augmentation with a benzodiazepine or conventional antipsychotic, either orally or intramuscularly
- Once daily dosing is possible
- Dosing twice daily can be used if desired for elderly people and children during dosage titration
- The only atypical antipsychotic with a liquid dosage formulation
- Tablets as small as 0.25 mg and 0.5 mg will be available soon
- New formulation that dissolves on the tongue is in preparation
- Depot palmitate formulation for monthly administration of the 9-hydroxy metabolite is in preparation
- Compliance monitoring, at point of care, for the presence of drug in saliva or urine will be available soon

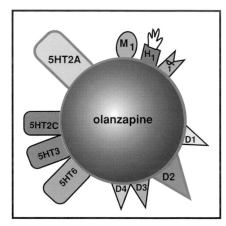

FIGURE 6.16 Olanzapine pharmacologic icon which portrays a qualitative consensus of current thinking about the binding properties of olanzapine. As for all atypical antipsychotic drugs discussed in this chapter, binding properties vary greatly with technique and species, and from one laboratory to another; they are constantly being revised and updated.

**Table 6.6
Olanzapine: Is it atypical?**

SDA?	Yes
Fewer EPS than with haloperidol?	Yes
Essentially patients never get EPS?	Yes and No[a]
Reduced incidence of tardive dyskinesia?	Probably
Less prolactin than with haloperidol?	Yes
Essentially prolactin never increases?	No[b]
Better negative symptoms than with placebo?	Yes
Better negative symptoms than with about 15 mg haloperidol?	Yes
Effective for symptoms refractory to conventional antipsychotics?	Maybe

[a]EPS unusual, except for occasional akathisia for doses of up to 15 mg/day, although EPS occasionally seen at doses above usual prescribing range (i.e. > 15 mg/day).
[b]Elevation of prolactin in fewer patients compared with haldoperidol and usually transient.

Table 6.7
Clinical pearls about olanzapine

- Well accepted for use in schizophrenia and bipolar disorders, including difficult cases
- Many anecdotal reports of use in children and treatment-refractory cases, and for positive symptoms of psychosis in disorders other than schizophrenia
- Documented efficacy as augmenting agent to SSRIs in nonpsychotic, treatment-resistant, major depressive disorders
- Muscarinic antagonist properties theoretically unfavorable for cognition in schizophrenia, and in elderly people, but cognitive symptoms of schizophrenia may improve in some clinical trials
- Same as placebo does not mean never, although EPS unusual at low to mid doses
- More weight gain than with other antipsychotics; this does not mean that every patient gains weight
- Not necessary to monitor liver function tests except in significant liver disease

Table 6.8
Dosage tips for olanzapine

- *More may be more:* By raising dose above recommended levels of 15 mg/day, it can be useful for acutely ill and agitated patients, and even for some treatment-resistant patients, gaining efficacy without many more side effects
- Thus, doses used in clinical practice (> 15 mg/day) are often *higher* than those suggested from clinical trials (i.e. usually 10 mg/day)
- Patients who respond to higher doses will incur higher drug costs, but this may be justified if the patient is severely ill and other options fail
- Higher doses given in the acute period or while agitated can sometimes be reduced later when the patient has been stabilized
- Rather than raising doses to a high level in difficult cases, consider augmentation with an oral or intramuscular benzodiazapine or conventional antipsychotic
- Women may require lower doses than men because plasma levels are higher in women than in men at comparable doses
- Once daily administration for all applications
- Four dosages: 2.5 mg, 5.0 mg, 7.5 mg and 10 mg
- Tablets are not scored, so they cannot be broken in half
- Each tablet can cost about the same, so a single dose size will allow 2.5, 5, 7.5 or 10 mg/day; the price can double for two dosages (required for 12.5, 15, 17.5 or 20 mg/day); it can triple for three dosages (required for 22.5, 25, 27.5 or 30 mg/day)
- A 15 -mg and a 20-mg tablet may be made available to reduce drug costs
- An intramuscular dosage formulation for acute administration is being developed

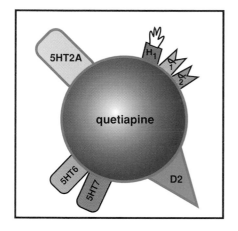

FIGURE 6.17   Quetiapine pharmacologic icon, which portrays a qualitative consensus of current thinking about the binding properties of quetiapine. As for all atypical antipsychotics, discussed in this chapter, binding properties vary greatly with technique and species, and from one laboratory to another; they are constantly being revised and updated.

Table 6.9
Quetiapine: Is it atypical?

SDA?	Yes
Fewer EPS than with haloperidol?	Yes
Essentially patients never get EPS?	Yes
Reduced incidence of tardive dyskinesia?	Expected[a]
Less prolactin than with haloperidol?	Yes
Essentially prolactin never increases?	Yes
Better negative symptoms than with placebo?	Yes
Better negative symptoms than with 12 mg haloperidol?	No[b]
Effective for symptoms refractory to conventional antipsychotics?	Maybe

[a]Studies in progress.
[b]The lower the dose of haloperidol, and the fewer EPS caused by an atypical antipsychotic, the harder it may be to beat haloperidol because of the lack of haloperidol-induced negative symptoms at low doses (see Tables 6.3 and 6.6).

Table 6.10
Clinical pearls about quetiapine

- Some patients respond to quetiapine who have failed to respond to other atypical antisychotics
- Anecdotal reports of usefulness for bipolar and treatment-refractory cases, and for positive symptoms of psychosis in disorders other than schizophrenia
- Early studies support use in adolescents and elderly people, and for hostility/aggression, cognition and affective symptoms in schizophrenia
- Better than placebo (but not haloperidol) for negative symptoms, although least likely to cause negative symptoms secondary to EPS
- May be the preferred antipsychotic for psychosis in Parkinson's Disease
- Never say never, but essentially no EPS or prolactin elevation at any dose
- Cataracts caused at high doses in dogs but not monkeys or humans, possibly as a result of species-specific inhibition of cholesterol biosynthesis in the lens of dogs
- However, in the USA, there is an FDA precaution to monitor for development of cataracts every 6 months (similar to precaution for carbamazepine and HMGCoA (hydroxymethylglutaryl-lCoA) reductase inhibitors of cholesterol biosynthesis)
- Postmarketing experience to date does not support a causal link between quetiapine and lens opacities
- Not necessary to have eyes examined until dose has been stabilized and plans made for long-term use

Table 6.11
Dosage tips for quetiapine

- Clinical trials suggest that a safe and effective dose range is 75 to 375 mg (US) or 75 to 400 mg (UK) twice daily for schizophrenia (except lower doses in elderly people, namely 25–75 mg twice daily)
- Clinical practice suggests target doses of 150–200 mg twice daily for schizophrenia
- Some clinical trials dosed three times daily, but, in clinical practice, doses are given twice daily, and may even change to once daily, especially for total daily doses of ≤ 400 mg, and after the patient has been stabilized for long-term treatment
- *More may be more:* By raising the dose, quetiapine can be useful for acutely ill and agitated patients, and even for some treatment-resistant patients, gaining efficacy essentially with no more side effects, especially for > 200 mg twice daily
- Three dosages: 25, 100 and 200 mg
- Tablets are not scored so they cannot be broken in half
- Recommended titration to 300–400 mg/day by day 4 requires two doses per day and changing combinations of 25 mg, 100 mg and 200 mg tablets
- In practice, aim for 200 mg twice daily on day 5, whether initiating new patients or switching while cross-titrating down on another antipsychotic, e.g. 50 mg on day 1 (25 mg twice daily or 50 mg qhs); 100 mg on day 2 (50 mg twice daily); 200 mg on day 3 (100 mg twice daily); 300 mg on day 4 (100 mg in the morning and 200 mg qhs) and 400 mg on day 5 (200 mg twice daily). Uses four 25 mg tablets, four 100 mg tablets, one 200 mg tablet in the first 4 days, then two 200 mg tablets daily thereafter
- Starter pack available in some countries
- Some patients may tolerate a simpler titration starting with 100 mg qhs on day 1; 100 mg bid on day 2; and 100 mg in the day and 200 mg at night on day 3. Can increase to 200 mg bid on day 4 or later, if inadequate response to 100 mg plus 200 mg
- Dosing in 100 mg increments simplifies dosing regimen and keeps costs down
- At 200 mg twice daily, quetiapine may be a lower cost atypical antipsychotic
- Doses higher or lower than 200 mg twice daily may be more expensive and require complicated combinations of multiple tablets and multiple dosage sizes (up to 10 tablets/day and all three dosage strengths)

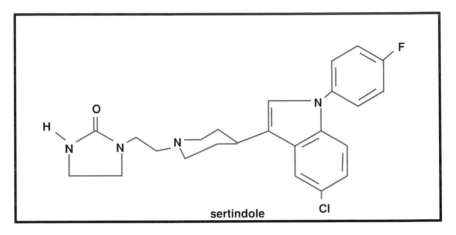

sertindole

FIGURE 6.18   Structure of sertindole. The chemical structure is related to that of serotonin.

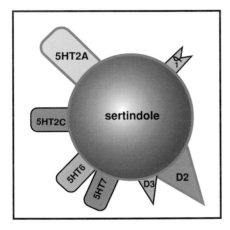

FIGURE 6.19   Sertindole pharmacologic icon, which portrays a qualitative consensus of current thinking about the binding properties of sertindole. As for all atypical antipsychotics discussed in this chapter, binding properties vary greatly with technique, and species, and from one laboratory to another; they are constantly being revised and updated.

Table 6.12
Sertindole: Is it atypical?

SDA?	Yes
Fewer EPS than with haloperidol?	Yes
Essentially patients never get EPS?	Yes
Reduced incidence of tardive dyskinesia?	Probably
Less prolactin than with haloperidol?	Yes
Essentially prolactin never increases?	No*
Better negative symptoms than with placebo?	Yes
Negative symptoms better than with 4,8 or 16 mg haloperidol?	Yes
Effective for symptoms refractory to conventional antipsychotics?	Maybe

[a]Elevations of prolactin in fewer patients compared with haloperidol; usually transient.

Table 6.14
Dosage tips for sertindole

- Start at 4 mg/day, increasing by 4 mg every 2–3 days with target dose 12–20 mg/day
- Some patients require 24 mg/day
- Tablets of 4, 12, 16 and 20 mg are available
- Once daily dosing

Table 6.13
Clinical pearls about sertindole

- Prolongs QTc on EKG
- Currently 'on hold' from marketing until relationship of QTc prolongation to ventricular arrhythmias and sudden deaths clarified
- Before withdrawal from marketing, pre-treatment EKG was recommended
- Contraindicated in patients with prolonged QT intervals, with significant cardiac disease, uncorrected hypokalemia, and concomitant use of other drugs that prolong the QT interval
- Accepted as efficacious for treatment of schizophrenia and bipolar disorders
- Anecdotal reports of use in children, and treatment-refractory cases, and for positive symptoms of psychosis in disorders other than schizophrenia
- May cause less weight gain than other antipsychotics
- Never say never, but essentially no EPS at any dose
- Can cause nasal congestion and decreased ejaculatory volume

Table 6.15
QT prolongation and antipsychotics

- Normal QT interval is generally less than 450–500 ms
- QT corrected for heart rate is called QTc (bradycardia prolongs QT)
- QT interval varies continuously within individuals, with values varying as much as 75–95 ms
- QT increases with food, sleep and athletic training
- Significant drug-induced prolongation of the QT interval may lead to sudden death caused by a lethal ventricular arrhythmia, *torsade de pointes* (literally, 'twisting of the points,' a continuously changing QRS vector)
- As a result of constant fluctuations in QT, a normal EKG does not rule out possible drug-induced QT prolongation at all times
- Psychotropic drugs such as thioridazine and amitriptyline may increase mean QT interval by 20 ms
- Sertindole increases mean QT by about 20 ms
- New atypical antipsychotics, including ziprasidone, risperidone, olanzapine and quetiapine, may all increase mean QT interval by 2–3 ms
- Significance of small increases in QT is unknown, drug-induced prolongations of QT of 5-15 ms are probably clinically insignificant
- Whether small drug-induced increases in the mean QT of a population predicts that there will be rare individuals with dangerous prolongations and risk of *torsade de pointes* and sudden death is also unknown
- Clinical studies are underway to evaluate whether there is any increased risk of sudden death for patients taking any atypical antipychotic, and also whether there are any differences in risk among ziprasidone, risperidone, olanzapine and quetiapine, which cause little or no mean prolongation of QT interval
- Clinical studies are also evaluating the possible relationship of sudden deaths reported for patients taking sertindole to drug-induced QT prolongation

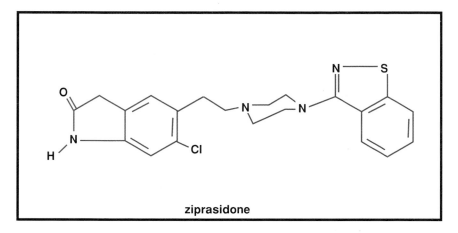

FIGURE 6.20   Structure of ziprasidone.

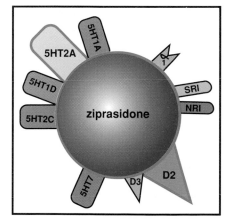

FIGURE 6.21   Ziprasidone pharmacologic icon which portrays a qualitative consensus of current theory about the binding properties of ziprasidone. As for all atypical antipsychotics discussed in this chapter, the binding properties vary greatly with technique, and species, and from one laboratory to another; they are constantly being revised and updated.

Table 6.16
Ziprasidone: Is it atypical?

SDA?	Yes
Fewer EPS than with haloperidol?	Yes
Essentially patients never get EPS?	Yes and No[a]
Reduced incidence of tardive dyskinesia?	Expected[b]
Less prolactin than with haloperidol?	Yes
Essentially prolactin never increases?	No[c]
Better negative symptoms than with placebo?	Yes
Better negative symptoms than with haloperidol?	Unknown[b]
Effective for symptoms refractory to conventional antipsychotics?	Unknown[b]

[a]EPS unusual except occasional akathisia.
[b]Studies in progress.
[c]Elevations of prolactin in fewer patients compared with haloperidol; usually transient.

Table 6.17
Clinical pearls about ziprasidone

- Little weight gain
- In addition to SDA pharmacology, unique binding properties include $5HT_{1A}$ agonist, $5HT_{1D}$ antagonist, $5HT_{2C}$ antagonist, and serotonin and norepinephrine reuptake blockade
- Unique pharmacologic profile suggests potential advantages for associated anxiety and depression
- Early study reporting use in children ages 7–17 with Gilles de la Tourete syndrome in which it is well tolerated and effective
- Approved in Sweden, pending approval in the USA, and currently not marketed anywhere
- Somnolence and dizziness are the most frequent side effects
- Prolongs the mean QT interval slightly, with greater prolongations in some individuals
- Significance of QT prolongations currently under investigation and will be resolved before approval in the USA

Table 6.18
Dosage tips for ziprasidone

- Very easy to dose (only two doses, 40 mg twice daily or 80 mg twice daily)
- No dosage adjustment for elderly people
- Recommended that it is taken with food, which can double ziprasidone's bioavailability
- May need to increase dose or beware of relapse if stabilized patients stop taking with food
- An intramuscular acute dosage formulation is being developed

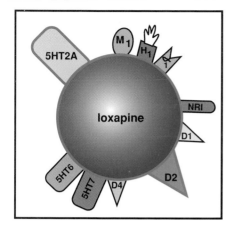

FIGURE 6.22   Loxapine pharmacologic icon. This icon portrays a qualitative consensus of current notions of the binding properties of loxapine. As for all antipsychotic drugs discussed in this chapter, binding properties vary greatly with technique, species and from one laboratory to another, and are constantly being revised and updated.

Table 6.19
Loxapine: Is it atypical?

SDA?	Yes
Fewer EPS than with haloperidol?	Maybe[a]
Essentially patients never get EPS?	No[b]
Reduced incidence of tardive dyskinesia?	No[b]
Less prolactin than with haloperidol?	No[b]
Better negative symptoms than with placebo?	Maybe[c]
Better negative symptoms than with haloperidol?	Maybe[c]
Effective for symptoms refractory to conventional antipsychotics?	Maybe[d]

[a]Some but not all studies show lower EPS than with haloperidol.
[b]EPS and elevated prolactin at doses of ≥ 60 mg/day; no studies of < 20 mg/day.
[c]No proper study of negative symptoms, but some studies show lower negative symptom-related items than conventional antipsychotics.
[d]No monotherapy study, but improved symptoms unresponsive to clozapine when added as an augmenting agent.

Table 6.20
Clinical pearls about loxapine

- Recently discovered to be a serotonin–dopamine antagonist (SDA) (binding studies and PET scans)

- Principal metabolites are also SDAs

- Developed as a conventional antipsychotic, i.e. reduces positive symptoms, but causes EPS and elevations of prolactin

- Lower EPS than with haloperidol in some studies, but no fixed dose studies and no low-dose studies

- Causes less weight gain than other antipsychotics, both atypical and conventional, and may even be associated with weight loss

- No formal studies of negative symptoms, but some studies show superiority to conventional antipsychotics for emotional withdrawal and social competence

- Norepinephrine reuptake inhibition (NRI; see Figure 6.22) of loxapine and metabolites suggests possible efficacy for depression

- Best use may be as a low-cost augmentation agent to atypical antipsychotics, e.g. enhances efficacy in clozapine partial responders when given together with clozapine

- For previously stabilized patients with 'breakthrough' agitation or incipient decompensation, can 'top-up' the atypical antipsychotic with intramuscular or oral single doses of loxapine as required

- For the 15% of patients maintained on two antipsychotics, can use loxapine to augment one atypical

- For the more than one-third of patients on doses of an atypical antipsychotic that are above the documented atypical/cost-effective ranges, consider lowering atypical antipsychotic dose and augmenting with loxapine 5–60 mg/day

- Loxapine costs only 10–25% of an atypical antipsychotic

Table 6.21
Dosage tips for loxapine

- Conventional antipsychotic properties at originally recommended doses (i.e. starting at 10 mg twice daily, maintenance 60–100 mg/day, maximum 250 mg/day given twice daily)

- Binding studies, PET studies and anecdotal clinical observations suggest that loxapine may be atypical at lower doses (perhaps 5–60 mg/day) but further studies are needed

- Anecdotal evidence that many patients can be maintained on 20–60 mg/day loxapine as monotherapy

- Available as 5 mg and 10 mg capsules for low-dose use and as 25 mg and 50 mg capsules for routine use

- Available as a liquid dosage formulation

- Only SDA available for acute intramuscular administration (50 mg/ml)

- Intramuscular loxapine may have faster onset of action and superior efficacy for agitated/excited and aggressive behavior than intramuscular haloperidol

- Give 25–50 mg i.m. (0.5–1.0 ml of 50 mg/ml solution) with onset of action within 60 min in the acute situation

- When initiating therapy with an atypical antipsychotic in an acute situation, consider intramuscular loxapine to 'lead in' to orally administered atypical antipsychotics, e.g. initiate oral dosing of an atypical antipsychotic with 25–50 mg loxapine two to three times daily short term as necessary for antipsychotic effects without EPS and sedation

- When using loxapine to 'top-up' previously stabilized patients now decompensating, use loxapine as single 25–50 mg doses as required intramuscularly or as oral liquid or tablets

- To augment partial responders to an atypical antipsychotic, consider doses of loxapine as low as 5–60 mg/day, but use full doses if necessary

Table 6.22
Weight gain and antipsychotics

No change or weight loss
Loxapine
Molindone

Increasing likelihood of weight gain

Ziprasidone (the least weight gain)
Thiothixene
Fluphenazine
Haloperidol
Risperidone
Chlorpromazine
Sertindole
Quetiapine
Thioridazine
Olanzapine
Zotepine
Clozapine (the greatest weight gain)

Weight is often ignored by clinicians, with body weight and body mass index not being monitored during long-term treatment. As many of the atypical antipsychotics in frequent use may increase body weight by 20–50 pounds with long-term maintenance, the selection of an atypical antipsychotic may be influenced by its actual or potential effects on weight gain. Regardless of which antipsychotic is chosen, the monitoring of body weight and prevention of obesity should be a priority for antipsychotic prescribers. New data document better than ever the increased health risks of obesity, such as new-onset diabetes mellitus, and especially accelerated cardiovascular mortality. This may be particularly relevant for the treatment of schizophrenic patients who are likely to have disproportionately high levels of other cardiovascular risk factors such as smoking (up to 80% of schizophrenics), sedentary lifestyle and unhealthy diets.

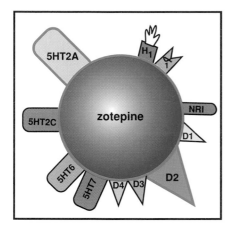

FIGURE 6.23   Zotepine pharmacologic icon, which portrays a qualitative consensus of current thinking about the binding properties of zotepine. As for all antipsychotics discussed in this chapter, binding properties vary greatly with technique and species, and from one laboratory to another; they are constantly being revised and updated.

Table 6.23
Zotepine: Is it atypical?

SDA?	Yes
Fewer EPS than with haloperidol?	Maybe[a]
Essentially patients never get EPS?	No
Reduced incidence of tardive dyskinesia?	Unknown
Less prolactin than with haloperidol?	No
Better negative symptoms than with placebo?	Unknown
Better negative symptoms than with haloperidol?	Maybe[b]
Effective for symptoms refractory to conventional antipsychotics?	Maybe

[a]Some but not all studies show lower EPS than conventional antipsychotics.
[b]Some studies show superiority to haloperidol, but were not placebo-controlled.

Table 6.24
Clinical pearls about zotepine

- Most common side effects are constipation, dry mouth, insomnia, sleepiness, asthenia and weight gain
- There is an active metabolite
- Marketed in Japan and Europe; in clinical trials in the USA
- Blocks norepinephrine reuptake as well as SDA properties, suggesting efficacy for depressive symptoms in schizophrenia

Table 6.25
Dosage tips for zotepine

- Target doses are 150–300 mg/day
- Increased risk of convulsions, especially over 300 mg/day
- Given three times daily

References

For further information about *clozapine*, see Casey 1998; Conley 1998; Glazer and Dickson 1998; Grohmann et al 1989; Honigfeld et al 1998; Huttunen 1995; Kane et al 1988; Lieberman 1998; Lieberman et al 1989 and 1998; MacGibbon et al 1994; Meltzer 1991; Meltzer 1994; Meltzer 1998; Meltzer et al 1998; Murray and Van Os 1998; Nordstrom et al 1993; Sharma and Mockler 1998; Sheitman et al 1998; Stahl 1998a; Stahl in press

For further informatiton about *risperidone*, see Aizenberg et al 1995; Almahfouz and Guay 1989; Aono et al 1978; Ataya et al 1988; Biller et al 1992; Bondolfi et al 1998; Bouloux and Grossman 1987; Byerly and DeVane 1996; Chouinard et al 1993; Chung and Eun 1998; Coccaro 1998; Conley et al 1998a; Coryell 1998; Daniel and Whitcomb 1998; Degen 1982; Doss 1979; Ereshefsky and Lacombe 1993; Farde et al 1995; Ghadirian et al 1982; Green et al 1997; Greenspan et al 1986; Greenspan et al 1989; Halbreich et al 1995; Halbreich and Palter 1996; Hillard 1998; Kee et al 1998; Kelly et al 1998; Kleinberg et al 1997; Klibanski et al 1980; Klibanski et al 1988; Klibanski and Greenspan 1986; Koppelman et al 1984; Lehmann 1986; Leonard et al 1989; Leysen et al 1994; Leysen et al 1998; Lindenmayer et al 1998; Lussier and Stip 1998; Marder et al 1997; Marder and Meibach 1994; Marken et al 1992; Masand 1998; Mitchell and Popkin 1982; Nelson et al 1998; Nestoros et al 1981; Netto and Claro 1993; Nyberg et al 1993; Nystrom et al 1988; Owens 1994;

Peuskens 1995; Ribot et al 1994; Rogers and Burke 1987; Schlecte 1995; Schlecte et al 1992; Schooler 1994; Schooler 1998; Schulz et al 1998; Segraves 1985; Segraves 1989; Sowers et al 1993; Stahl 1998a, 1998b and 1998c; Stahl in press; Tecott et al 1995; Tohen and Zarate 1998; Tracy et al 1998; Wardlaw and Bilezikian 1992; Windgassen et al 1996; Wirshing et al 1997; Zayas and Grossberg 1998

For further information about *olanzapine*, see Beasley et al 1996a and 1996b; Beasley et al 1997; Beuzen et al 1998; Bymaster et al 1996; Conley et al 1998a and 1998b; Crawford et al 1997; Gheuens and Grebb 1998; Glazer 1997; Hamilton et al 1998; Kasper and Kufferle 1998; Kinon et al 1998; Li et al 1998; Littrell et al 1998; Martin et al 1997; Nyberg et al 1997; Purdon 1998; Ring et al 1996; Robertson and Fibiger 1992; Robertson and Fibiger 1996; Schooler 1998; Shelton et al 1998; Stahl 1998a; Stahl in press; Stockton and Rasmussen 1996; Street et al 1996; Street et al 1998; Taylor et al 1998; Tollefson et al 1997a and 1997b; Tollefson et al 1998a and 1998b; Tollefson and Tran 1998a, 1998b and 1998c; Tran et al 1997a and 1997b; Zarate et al 1998

For further information about *quetiapine*, see Arvantis et al 1997; Arvanitis and Rak 1998; Borison et al 1996; Gefvert et al 1998; Goldstein 1998a, 1998b and 1998c; King et al 1998; McConville et al 1998; Meats 1997; Peuskens and Link 1997; Sax et al 1998; Small et al 1997; Smith et al 1997; Stahl 1998a; Stahl in press

For further information about *sertindole*, see Daniel et al 1998; Fritze and Bandelow 1998; Hale 1998; Kane 1997; Kasper et al 1998; Pilowsky et al 1997; VanKammen et al 1996; Zimbroff et al 1997

For further information about *ziprasidone*, see Allison et al 1998; Arato et al 1997a and 1997b; Aweeka et al 1997; Brook et al 1998; Chappell and Sallee 1998; Daniel et al 1997; Daniel et al in press; Everson et al 1997; Fischman et al 1996; Goff et al 1998; Gunn et al 1997; Keck et al 1998; Miceli et al 1997; Muirhead et al 1996; Prakash et al 1997a and 1997b; Reeves et al 1998a, 1998b, 1998c and 1998d; Schmidt et al 1998; Seeger et al 1995; Serper and Chou 1997; Sprouse et al 1998a and 1998b; Swift et al 1998a and 1998b; Tandon et al 1997; Tensfeldt et al 1997; Wilner et al 1996; Wilner et al 1997a and 1997b; Zorn et al 1998

For further information about *loxapine*, see Al-Jeshi et al 1996; Bishop et al 1977; Branchey et al 1978; Buckley in press; Carlyle et al 1993; Charalampous et al 1974; Cheung et al 1991; Cole et al 1982; Cohen et al 1982; Coupet et al 1976; Coupet et al 1979; Coupet and Rauh 1979; Dahl 1982; Dean and Gallant 1979; Deniker et al 1980; DePaulo and Ayd 1982; Dubin and Weiss 1986; Ereshefsky in press; Filho et al 1975; Fruensgaard and Jensen 1976; Fruensgaard et al 1977; Fulton et al 1982; Glazer in press; Hue et al 1998; Kapur et al 1996; Kapur et al 1997; Kiloh et al 1976; Kramer et al 1978; Lehmann et al 1981; Leone 1979; Meltzer and Jayathilake in press; Midha et al 1993; Mowerman and Siris 1996; Moyano 1975; O'Connell and Lieberman 1978; Paprocki et al 1976; Paprocki and Versiani 1977; Remington and Kapur in press; Richelson in press; Robertson et al 1982; Schiele 1975; Selkin 1979; Selman et al 1976; Serban 1979; Simpson et al 1978; Singh et al 1996; Stahl in press; Thomas 1979; Trichard et al 1998; Tuason 1986; Tuason et al 1984; Van der Velde and Kiltie 1975; Vanelle et al 1994; Zisook et al 1978

For further information about *zotepine*, see Kilpatrick et al 1998a and 1998b; Needham et al 1998; Palmgren et al 1998; Petit et al 1998; Prakash and Lamb 1998; Reynolds et al 1998; Welch et al 1998

For additional reading see Alexopoulos et al 1998; Bitello et al 1997; Buckley 1998; Dellenberg and Hopkins 1996; Deutsch et al 1992; Findling et al 1996; Keck et al 1994; Keck et al 1996; Marcus et al 1996; Marder 1998a and 1998b; Meltzer et al 1998; Moghaddam and Bunney 1990; Nguyen et al 1992; Pohen and Zarate 1998; Robertson et al 1994; Roth and Meltzer 1996; Roth et al 1995; Roth and Elsworth 1996; Sciolla and Jeste 1998; Seeman et al 1997; Stahl 1997; Sunderland 1996; Walsch 1998; Wan et al 1995; Weiss and Kilts 1998; Zayas and Grossberg 1998

CHAPTER 7

THE CYTOCHROME P450 SYSTEM AND PHARMACOKINETIC CONSIDERATIONS FOR ATYPICAL ANTIPSYCHOTICS

Recently, there has been a veritable explosion in the understanding of the pharmacokinetic basis of drug interactions. Recall that pharmacokinetics is the study of how the body acts upon drugs, especially to absorb, distribute, metabolize and then excrete them. Such pharmacokinetic actions are mediated predominantly through the hepatic drug-metabolizing system known commonly as cytochrome P450 enzyme systems.

The cytochrome P450 enzyme systems and the pharmacokinetic actions they represent must be contrasted with the pharmacodynamic actions of drugs. In fact, this entire book has dealt almost exclusively so far with the pharmacodynamics of antipsychotics, especially how atypical antipsychotics act upon the brain.

In the visual lessons that follow, the key enzymes that mediate important pharmacokinetic interactions with atypical antipsychotic drugs are shown. In addition, the consequences of these pharmacokinetic actions of atypical antipsychotics upon concomitantly administered drugs are shown. This will lead to obvious tips for avoiding unwanted drug interactions, based on an understanding of the pharmacokinetics of atypical antipsychotics and the cytochrome P450 enzyme systems.

Table 7.1
Pharmacokinetics How the body acts on drugs **Pharmacodynamics** How drugs act upon the body (especially the brain)

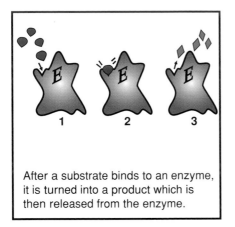

After a substrate binds to an enzyme, it is turned into a product which is then released from the enzyme.

FIGURE 7.1. A substrate in panel 1 interacts with an enzyme by binding to the enzyme's active site in panel 2, and is transformed into a product in panel 3.

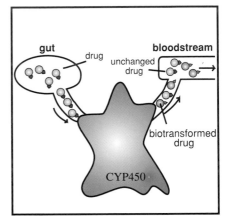

FIGURE 7.2. This figure shows how a drug is absorbed and delivered through the gut wall to the liver to be biotransformed so that it can eventually be excreted and eliminated from the body. Specifically, the P450 enzyme in the gut wall or liver converts the drug substrate into a biotransformed product in the bloodstream. After passing through the gut wall and liver, the drug will exist partially as unchanged drug and partially as transformed product.

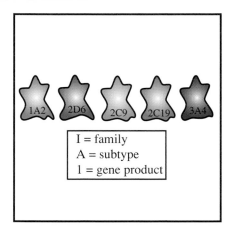

I = family
A = subtype
1 = gene product

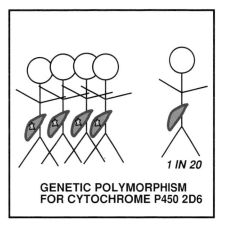

1 IN 20

GENETIC POLYMORPHISM
FOR CYTOCHROME P450 2D6

FIGURE 7.3. There are several known cytochrome P450 systems. Five of the most important enzymes for psychotropic drug metabolism are shown here. The first number is the cytochrome P450 family; the subtype is then indicated with a letter, followed by another number indicating the specific gene product, which that cytochrome P450 enzyme represents. There are over 30 known enzymes, and probably many more awaiting discovery and classification.

FIGURE 7.4. Not all individuals have all the same cytochrome P450 enzymes. In such cases, the enzyme is said to be polymorphic. For example, about 5-10% of Caucasians are poor metabolizers via the enzyme 2D6. They must metabolize drugs by alternative routes which may not be as efficient as the 2D6 route.

Another enzyme, 2C19, has reduced activity in approximately 20% of Japanese and Chinese individuals, and in 3-5% of Caucasians.

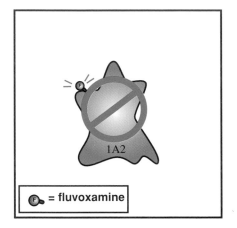

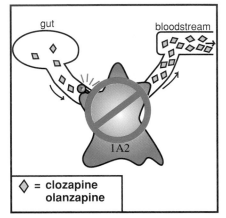

FIGURE 7.5. One cytochrome P450 enzyme of relevance to psychotropic drugs is 1A2. This enzyme is inhibited by the serotonin-selective reuptake inhibitor (SSRI) fluvoxamine.

FIGURE 7.6. When fluvoxamine is given together with other drugs that use 1A2 for their metabolism, those drugs can no longer be metabolized efficiently. An example of a potentially important interaction is when fluvoxamine is given together with either clozapine or olanzapine. In this case, the clozapine or olanzapine dose may need to be lowered or else the blood levels could rise, and possibly cause side effects.

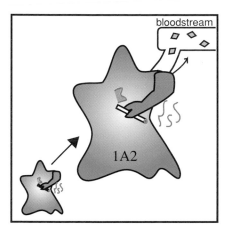

FIGURE 7.7. Cigarette smoking can induce 1A2, thereby increasing its enzyme activity. Thus, patients who start smoking or increase their level of smoking could have a decrease in their clozapine or olanzapine levels, risking the chance of a psychotic relapse. Also, cigarette smokers may require higher doses of these atypical antipsychotics than nonsmokers.

Table 7.2
Prevention of unwanted pharmacokinetic interactions. I: Tips for avoiding 1A2 interactions with atypical antipsychotics

Fluvoxamine may raise clozapine and olanzapine levels
So could large amounts of grapefruit juice and the antibiotics ciprofloxacin and norfloxacin
Cigarette smoking can lower clozapine and olanzapine levels

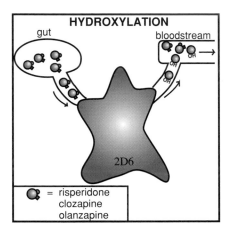

HYDROXYLATION

gut

bloodstream

2D6

= risperidone
clozapine
olanzapine

FIGURE 7.8. Another cytochrome P450 enzyme of importance to prescribers of atypical antipsychotic drugs is the enzyme 2D6. This enzyme hydroxylates various drugs, including risperidone, leading to 9-hydroxy-risperidone, which is active as an atypical antipsychotic just like risperidone itself. Clozapine and olanzapine are also substrates for 2D6, but their metabolites are not active.

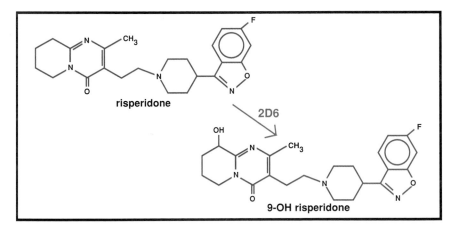

risperidone

2D6

9-OH risperidone

FIGURE 7.9. Metabolism of risperidone to an active metabolite, 9-hydroxy- risperidone, by the cytochrome P450 enzyme 2D6.

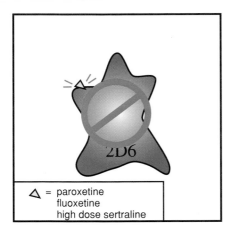

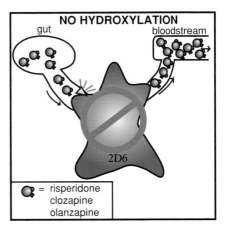

FIGURE 7.10. Several antidepressants are inhibitors of 2D6. This includes the SSRIs, especially paroxetine and fluoxetine, although individual SSRIs differ in the potency of this effect from one to another.

FIGURE 7.11. When risperidone, olanzapine or clozapine is given together with SSRIs (especially paroxetine, fluoxetine or high doses of sertraline), the plasma concentrations of these atypical antipsychotics can rise.

For risperidone, the clinical significance is uncertain, as both the parent and the metabolite are active.

For olanzapine and clozapine, concomitant administration with these SSRIs may require a lowering of the dose of the atypical antipsychotic if increased side effects become apparent.

Table 7.3
Prevention of unwanted pharmacokinetic interactions. II: Tips for avoiding 2D6 interactions

SSRIs (particularly paroxetine, fluoxetine and high doses of sertraline) can increase risperidone, clozapine and olanzapine levels

Interactions of SSRIs and atypical antipsychotics have, however, uncertain clinical significance

SSRIs can increase tricyclic antidepressant (TCA) levels which can lead to TCA toxicity

SSRIs can block the pain relief of codeine and raise codeine levels

SSRIs can increase β-blocker levels

Table 7.4
Prevention of unwanted pharmacokinetic interactions. III: Tips for avoiding 2C interactions

Fluvoxamine and fluoxetine can raise diazepam and phenytoin levels

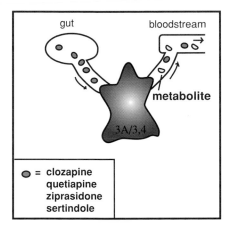

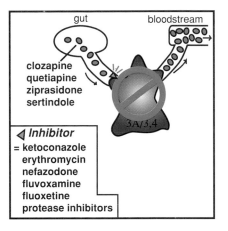

FIGURE 7.12. The cytochrome P450 enzyme 3A4 metabolizes four of the atypical antipsychotics, including clozapine, quetiapine, ziprasidone and sertindole. Several psychotropic drugs are weak inhibitors of this enzyme, including fluvoxamine, nefazodone and the active metabolite of fluoxetine, norfluoxetine. Several nonpsychotropic drugs are powerful inhibitors of 3A4, including ketoconazole (antifungal), protease inhibitors (for AIDS/HIV) and erythromycin (antibiotic).

FIGURE 7.13. For these four atypical antipsychotics, the clinical implication is that concomitant administration with a 3A4 inhibitor may require dosage reduction of the atypical antipsychotic. This may be especially important for drugs that are capable of prolonging the QTc interval (see Table 6.15); that is, it may be clinically insignificant that certain drugs prolong QTc interval at their usual plasma concentrations. However, if their plasma concentrations rise significantly as a result of 3A4 inhibition, this may cause clinically significant cardiac arrhythmias. This is exemplified by several nonpsychotropic drugs, including astemizole and cisapride. This may theoretically be the case for sertindole, although this remains to be proved. Nevertheless, it is prudent to monitor closely or avoid concomitant administration of 3A4 inhibitors with sertindole until further information on the safety of these combinations is available.

Therefore, on theoretical grounds, nefazodone, fluvoxamine and perhaps even other SSRIs should be cautiously administered with sertindole, as well as ketoconazole, protease inhibitors and erythromycin.

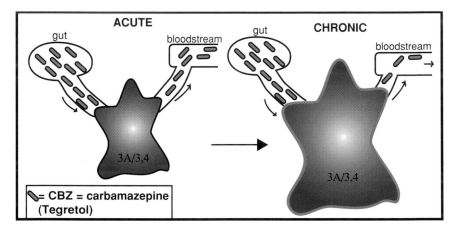

FIGURE 7.14.   Drugs can not only be substrates for a cytochrome P450 enzyme, or an inhibitor of a P450 enzyme, they can also be inducers of a cytochrome P450 enzyme and thereby increase the activity of that enzyme. One example of this is carbamazepine which is both a substrate and an inducer of 3A4. Thus, when first starting patients on carbamazepine, the levels of carbamazepine in the plasma will reflect the activity of the baseline, noninduced form of 3A4 (left). Over time, however, the activity of 3A4 increases as a result of induction of this enzyme by carbamazepine. This can lead to reduction of carbamazepine levels, because the increased enzyme activity leads to more efficient metabolism of the substrate carbamazepine (right). To keep carbamazepine plasma levels constant, the dose may have to be increased to adjust for the autoinduction of 3A4.

The same consideration must be made when carbamazepine is added to the regimen of patients receiving clozapine, quetiapine, ziprasidone or sertindole; that is, the doses of these atypical antipsychotics may need to be increased.

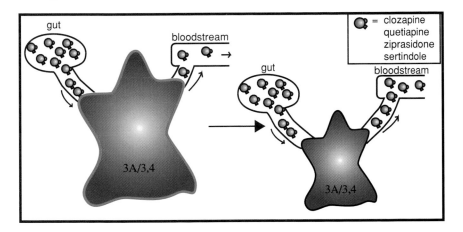

FIGURE 7.15. If carbamazepine is stopped in a patient receiving clozapine, quetiapine, ziprasidone or sertindole, the doses of these antipsychotics may need to be reduced, because the autoinduction of 3A4 by carbamazepine will reverse over time (see also Figure 7.14).

Table 7.5
Prevention of unwanted pharmacokinetic interactions. IV: Tips for avoiding 3A3/4 interactions

In terms of nonpsychotropic drugs, addition of astemizole or cisapride to erythromycin, ketoconazole or protease inhibitors could be a dangerous or even lethal combination

In terms of atypical antipsychotics, dose reduction of 3A4 substrates, such as clozapine, quetiapine, ziprasidone and sertindole, may be required when added to powerful 3A4 inhibitors, such as erythromycin, ketoconazole or protease inhibitors

This may be especially important for sertindole or any other atypical antipsychotic which prolongs QTc interval

On theoretical grounds, dose reduction of the 3A4 substrates clozapine, quetiapine, ziprasidone and sertindole may even be required when administered together with weak 3A4 inhibitors such as nefazodone, fluvoxamine or fluoxetine, especially if such combinations cause side effects

Table 7.6
Pharmacokinetic pearls about clozapine

Inhibitors of 1A2 (for example, fluvoxamine) can raise clozapine levels

Cigarette smoking (an inducer of 1A2) can decrease clozapine levels

Be vigilant for relapse if patient starts or increases smoking

Inhibitors of 2D6 (for example, some SSRIs such as fluoxetine, paroxetine and high doses of sertraline) can also raise clozapine levels

Strong inhibitors of 3A4 (for example, erythromycin, ketoconazole and protease inhibitors) can also raise clozapine levels

Weak inhibitors of 3A4 (for example, nefazodone, fluvoxamine and fluoxetine) may theoretically raise clozapine levels

Plasma half-life suggests twice daily administration, but, in practice, it may be given once a day at night

**Table 7.7
Pharmacokinetic pearls
about risperidone**

May have the fewest clinically
relevant drug interactions of the
atypical antipsychotics
Substrate for the cytochrome
P450 enzyme 2D6 which
converts it to an active
metabolite
If given with 2D6 inhibitors (for
example, some SSRIs such as
fluoxetine, paroxetine and high
doses of sertraline), net
efficacy unchanged
Plasma half-life suggests twice
daily administration, but clinical
experience suggests once daily
is sufficient for efficacy,
especially if total daily dose 4
mg or less
A long-acting depot formulation
of 9-hydroxyrisperidone is being
developed
A liquid dosage formulation is
available

**Table 7.8
Pharmacokinetic pearls
about olanzapine**

Inhibitors of 1A2 (for example,
fluvoxamine) can raise
olanzapine levels
Cigarette smoking (an inducer of
1A2) can decrease olanzapine
levels
Inhibitors of 2D6 (for example,
some SSRIs such as
fluoxetine, paroxetine and high
doses of sertraline) may raise
olanzapine levels
In practice, olanzapine doses do
not generally need to be
adjusted in smokers or in
patients taking SSRIs
Well documented to have
efficacy with once daily
administration as a result of its
long half-life
Women may have higher plasma
drug levels than men, and thus
require lower doses
An acute acting intramuscular
dosage formulation is being
developed

Table 7.9
Pharmacokinetic pearls about quetiapine

Short plasma half-life (3–4 hours)

Clinical experience suggests twice daily dosing is sufficient for efficacy

Strong inhibitors of 3A4 (for example, erythromycin, ketoconazole and protease inhibitors) may raise quetiapine levels

Weak inhibitors of 3A4 (for example, nefazodone, fluvoxamine and fluoxetine) may theoretically raise quetiapine levels

Phenytoin can lower quetiapine levels, necessitating a dosage increase for quetiapine

Table 7.10
Pharmacokinetic pearls about sertindole

Long half-life (3 days), so once daily dosing

Strong inhibitors of 3A4 (for example, erythromycin, ketoconazole and protease inhibitors) may raise sertindole levels

Weak inhibitors of 3A4 (for example, nefazodone, fluvoxamine and fluoxetine) could theoretically raise sertindole levels

Caution advised to avoid or monitor closely concomitant administration with 3A4 inhibitors, which could theoretically cause cardiotoxicity if they cause large increases in plasma sertindole levels

Weakly inhibits 3A4 itself

Is a 2D6 inhibitor, so use with caution with tricyclic antidepressants, because sertindole could raise their levels

Carbamazepine and phenytoin decrease sertindole levels and may require an increase in sertindole dosage

2D6 and 3A4 inhibitors reduce sertindole clearance and may increase its levels; thus, concomitant use of fluoxetine, paroxetine, erythromycin and calcium channel blockers should be done with caution; concomitant administration of ketoconazole and other potent 3A4 inhibitors is contraindicated

E

Table 7.11
Pharmacokinetic pearls about ziprasidone

Short half-life (6 hours), but duration of binding from PET studies long enough for twice daily dosing

Strong inhibitors of 3A4 (for example, erythromycin, ketoconazole and protease inhibitors) may raise ziprasidone levels, but dosage adjustment probably not necessary

Weak inhibitors of 3A4 (for example, nefazodone, fluvoxamine and fluoxetine) may also raise ziprasidone levels

An acute acting intramuscular dosage formulation is at an advanced stage of research

No dosage adjustment necessary for elderly people

Should be taken with food, which can double ziprasidone's bioavailability

Table 7.12
Pharmacokinetic pearls about zotepine

Given three times daily (half-life of 8 hours)

Active metabolite formed by N-demethylation

Plasma concentrations elevated in elderly people

Plasma concentrations are raised by benzodiazepines

It is unclear which CYP450 enzymes are involved in zotepine's metabolism

Table 7.13
Pharmacokinetic pearls about loxapine

Major metabolite is 8-hydroxy-loxapine, a weak SDA with norepinephrine reuptake properties as well

Secondary metabolite is 7-hydroxy-loxapine, a very potent SDA which also has norepinephrine reuptake blocking properties

More 8-hydroxy-loxapine is present than loxapine itself in the blood after oral administration

More loxapine than 8-hydroxy-loxapine is found in the blood after intramuscular administration

Also metabolized to amoxapine, a known antidepressant, in low amounts

Significant amounts of 8-hydroxy-amoxapine are formed

Small amounts of the 7-hydroxy metabolite of amoxapine are also formed

7-Hydroxy-amoxapine is a very potent SDA and 8-hydroxy-amoxapine is a weak SDA

Amoxapine and its 7- and 8-hydroxy metabolites are all norepinephrine reuptake inhibitors

Pharmacology of loxapine and its metabolites suggests potential antidepressant activity

It is the only SDA currently available in an intramuscular dosage formulation for acute administration

Pattern of metabolites after intramuscular administration suggests a more potent net SDA than after oral administration

Available as a liquid dosage formulation

Oral dosing is twice daily as monotherapy, but may be used once or twice daily when augmenting an atypical antipsychotic, especially at lower doses

Table 7.14
Pharmacokinetics: summary

Many drug interactions are primarily of academic or marketing interest

Many drug interactions are statistically significant yet clinically insignificant or easily managed by dosage adjustments

Few combinations must be absolutely avoided

Several combinations require dosage adjustment of one of the drugs

References

See the end of Chapter 6 for references and further information on clozapine, risperidone, olanzapine, quetiapine, sertindole, ziprasidone, loxapine and zotepine.

CHAPTER 8

ANTIPSYCHOTIC SUMMARY

In this book, we have reviewed the pharmacology of four neurotransmitter systems which are key to understanding the mechanism of action of antipsychotic drugs, namely dopamine, serotonin, acetylcholine and glutamate (see Chapter 2). We have also explored the five dimensions of symptoms in schizophrenia, in order to explain the potential scope of therapeutic action of the antipsychotics. These include the positive, negative, cognitive, affective and aggressive/hostile symptoms of schizophrenia and related illnesses (see Chapter 3). This background then allowed the psychopharmacology of over a dozen antipsychotic to be characterized as conventional antipsychotics, serotonin–dopamine antagonists (SDAs), and atypical antipsychotics (see Chapters 4 and 5).

Eight antipsychotics have been profiled in detail (see Chapters 6 and 7). Three of these drugs are atypical antipsychotics which are now considered to be first-line therapy for schizophrenia, namely risperidone, olanzapine and quetiapine. Conventional antipsychotics have been pushed into second-line or supplemental use. Clozapine originally defined the concept of an atypical antipsychotic, but is relegated to second-line use because of its potential bone marrow toxicity and consequent need for frequent monitoring of blood counts. Sertindole is an atypical antipsychotic originally introduced into clinical practice but then withdrawn from marketing while further testing of its cardiac safety is carried out. Ziprasidone is currently awaiting regulatory approval in many countries, and is poised to join the other three atypical antipsychotics as a first-line treatment choice. Both loxapine and zotepine are serotonin–dopamine antagonists with interesting clinical profiles, but they cannot currently be considered to meet the standard of a first-line atypical antipsychotic.

Clinical pearls and dosage tips were given for each of these eight antipsychotic drugs to assist the reader in choosing which antipsychotic to prescribe for which patient (see Chapter 6). In addition, pharmacokinetic pearls were given for these eight drugs to help anticipate and manage drug interactions (see Chapter 7). Here, in Chapter 8, we summarize the use of antipsychotics for treating the five symptom dimensions in schizophrenia and related illnesses, and list some overall clinical generalizations about the use of atypical antipsychotics in clinical practice.

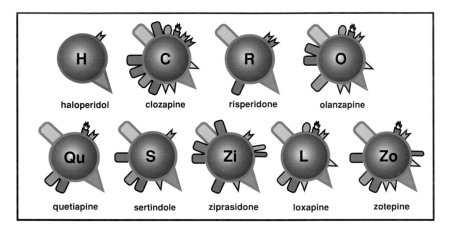

FIGURE 8.1.   Pharmacologic icons for nine antipsychotic drugs: haloperidol (H), clozapine (C), risperidone (R), olanzapine (O), quetiapine (Qu), sertindole (S), ziprasidone (Zi), loxapine (L) and zotepine (Zo). Ziprasidone is still in clinical development, zotepine is not available in all countries and sertindole is currently on hold from marketing at the time of writing. Thus, current therapeutics for psychosis emphasize three atypical antipsychotics, which are considered to be first-line agents worldwide, namely risperidone, olanzapine and quetiapine. Clozapine is generally the second-line treatment after first-line treatments fail. Loxapine is not an atypical antipsychotic, but may be useful in patients who require augmentation of therapeutic response.

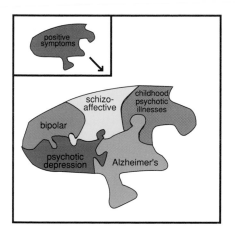

FIGURE 8.2.   Positive symptoms of psychosis. These were discussed in Chapter 3 (see Figure 3.3 and Table 3.1). Schizophrenia is not the only disorder associated with positive symptoms, because Alzheimer's disease, psychotic depression, bipolar disorders, schizoaffective disorder and various childhood psychotic disorders can also have associated positive psychotic symptoms.

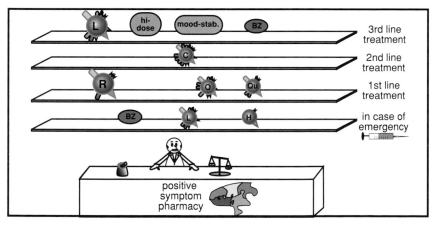

FIGURE 8.3.   Treatment of positive symptoms of psychosis. The atypical antipsychotics risperidone (R), olanzapine (O) and quetiapine (Qu) are considered to be the first-line treatment for disorders characterized by positive symptoms of psychosis. As there are as yet no injectable formulations of any atypical antipsychotic available, in an emergency situation injections of conventional antipsychotics such as loxapine (L) or haloperidol (H), or a sedating benzodiazepine such as lorazepam (BZ), can provide a quick treatment option for patients. If all the first-line atypical antipsychotics fail to provide adequate relief of positive symptoms, clozapine is a possible second-line treatment. Other options, if first-line treatments fail, are to try higher than normal doses of first-line treatments, or to augment with a conventional antipsychotic (especially low-dose loxapine), a sedating benzodiazepine (especially lorazepam) or a mood stabilizer (especially valproic acid).

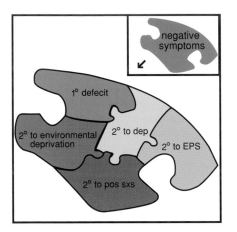

FIGURE 8.4. Negative symptoms of psychosis can be either primary or secondary. These symptoms were discussed in Chapter 3 (see Figure 3.5 and Tables 3.2 and 3.3). Primary negative symptoms are considered to be those that are core to primary deficits of schizophrenia itself. This may also include negative symptoms associated with or thought to be secondary to the positive symptoms of psychosis. Other secondary negative symptoms are those associated with EPS, especially those EPS caused by antipsychotics. Other secondary negative symptoms are those associated with depressive symptoms and with environmental deprivation.

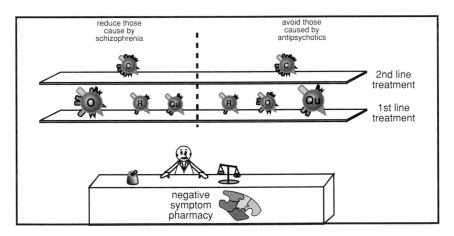

FIGURE 8.5. Treatment of negative symptoms of psychosis. The atypical antipsychotics olanzapine (O), risperidone (R) and quetiapine (Qu) are all considered to be first-line treatments for the primary symptoms of schizophrenia. Qu, O and R are also likely to avoid the worsening of negative symptoms because they do not generally cause EPS.

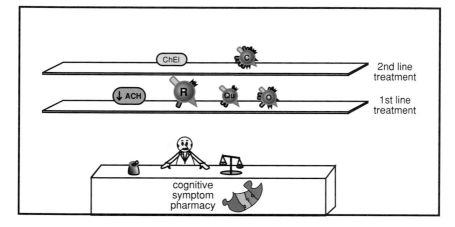

FIGURE 8.6.   Cognitive symptoms of schizophrenia were discussed in Chapter 3 (see Figure 3.6 and Tables 3.4 and 3.5). New data are emerging which suggest that all the first-line atypical antipsychotics, including risperidone (R), quetiapine (Qu) and olanzapine (O), can improve cognitive functioning in some patients. It is also important to reduce anticholinergic treatments if cognitive dysfunction is a problem and the patient is taking drugs with anticholinergic properties. Second-line treatments for cognitive disorders in schizophrenia can include clozapine, or cholinesterase inhibitors (ChEI) such as donepezil.

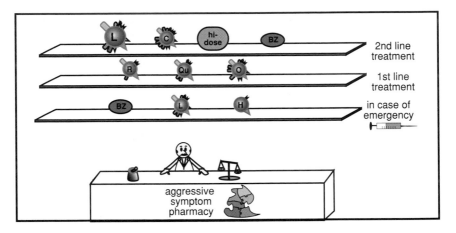

FIGURE 8.7. Hostility and aggressiveness in schizophrenia were discussed in Chapter 3 (see Figure 3.7 and Table 3.6). The atypical antipsychotics risperidone (R), quetiapine (Qu) and olanzapine (O) all are first-line treatments for these symptoms. In an emergency situation, injections of a sedating benzodiazepine (BZ), such as lorazepam, or a conventional antipsychotic, such as loxapine (L) or haloperidol (H), may be necessary for uncontrolled hostility, aggression and violent behavior. Clozapine (C) can be a particularly powerful second-line treatment for inadequately controlled aggression and violent behavior when the three first-line atypical antipsychotics give an unsatisfactory therapeutic response. Other second-line treatments when atypical antipsychotics fail are use of high doses of an atypical antipsychotic, or augmentation with loxapine (L) or a sedating benzodiazepine (BZ) such as lorazepam.

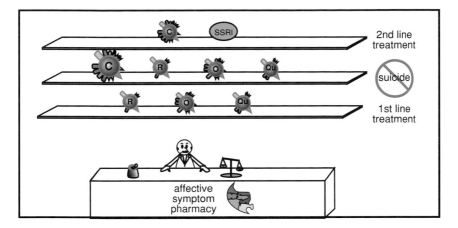

FIGURE 8.8.   Affective symptoms such as depression and anxiety can be associated with schizophrenia, even when they do not reach full diagnostic criteria for a DSM-IV disorder (see Chapter 3, Figure 3.8 and Table 3.7). New data from the three first-line atypical antipsychotics, including risperidone (R), olanzapine (O) and quetiapine (Qu), suggest that each of them can reduce such symptoms in schizophrenia. Clozapine (C) is documented as reducing suicide in schizophrenia, although the three first-line atypical antipschotics may also do this. Clozapine is also a second-line treatment for affective disorders when first-line atypical antipsychotics are not satisfactory. Augmentation with an antidepressant such as an SSRI (serotonin selective reuptake inhibitor) may also be considered if there is no satisfactory response to an atypical antipsychotic alone.

Table 8.1
Atypical antipsychotics: Where clinical practice confirms clinical trials

Atypical antipsychotics undoubtedly have reduced EPS compared with conventional antipsychotics

Atypical antipsychotics probably reduce negative symptoms of schizophrenia better than the conventional antipsychotics, but this may in part be secondary to reduced EPS

Atypical antipsychotics possibly reduce cognitive and affective symptoms in schizophrenia, which in part may also be secondary to reduced EPS

The magnitude of these properties makes atypical antipsychotics the first-line therapy for psychosis, and conventional antipsychotics the second-line

Table 8.2
Atypical antipsychotics: Where clinical practice differs from clinical trials

Different atypical antipsychotics often have clinically distinctive effects in different patients

Optimal doses derived from clinical trials do not match optimal doses used in clinical practice

Atypical antipsychotics do not always seem to work as fast as conventional antipsychotics

Atypical antipsychotics can appear to be less effective than conventional antipsychotics in treating acute psychosis, especially in the first few days

Atypical antipsychotics can appear to be less effective than conventional antipsychotics in treating agitation (especially if acute)

Efficacy of new atypical antipsychotics in treating patients refractory to conventional antipsychotics is not dramatic; clozapine is still the gold standard

To switch from ongoing treatment to a new drug, some clinical trials recommended 'stop–start' whereas, in practice, patients are usually 'cross-tapered' (i.e. the immediate discontinuation of ongoing treatment followed by the immediate start of a new atypical drug can lead to rebound psychosis, withdrawal reactions and rehospitalizations, so the dose of ongoing treatment can be tapered down, while simultaneously tapering up the new atypical)

Monotherapy with an atypical antipsychotic is considered to be inadequate in up to 20% of patients who also receive augmentation with a second antipsychotic

Occasional patients may actually respond better to conventional antipsychotics than to atypical antipsychotics

Table 8.3
Managing inadequate treatment responses to antipsychotics

The atypical antipsychotics risperidone, olanzapine and quetiapine are first-line treatments

If one of these three agents generates an unsatisfactory treatment response at normal doses, try one of the other atypical antipsychotics

If the second drug is also unsatisfactory, try the third

If all three agents are unsatisfactory, consider higher doses than usual, a trial of clozapine or various augmentation strategies

Augmentation can be with a conventional antipsychotic such as loxapine or haloperidol, with a benzodiazepine such as lorazepam or with a mood stabilizer such as valproic acid

REFERENCES

Aghajanian GK (1996) Electrophysiology of serotonin receptor subtypes and signal transduction pathways. In: Bloom FE, Kupfer DJ, eds, *Psychopharmacology: the Fourth Generation of Progress* (Raven Press: New York) 451–60

Aizenberg D, Zemishlany Z, Dorfman-Etrog P, Weizman A (1995) Sexual dysfunction in male schizophrenic patient, *J Clin Psychiatry* 56:137–41

Al-Jeshi A, Jeffries JJ, Kapur S (1996) Loxapine: an enigma, *Can J Psychiatry* 41(2):131–2

Alexopoulos GS, Silver JM, Khan DA, et al (1998) Treatment of agitation in older patients with dementia, *Postgrad Med* special report, April:1–88

Allison DB, Mentore JL, Heo M et al (1998) Weight gain associated with conventional and newer antipsychotics: a meta-analysis, *Eur Psychiatry* 13(4):302s

Almahfouz A, Guay AT (1989) Hyperprolactinaemia and impotence: why, when and how to investigate [Letter], *J Urol* 142:1080.

Aono T, Shioji T, Kinugasa T et al (1978) Clinical and endocrinological analyses of patients with galactorrhea and menstrual disorders due to sulpiride or metroclopramide, *J Clin Endocrinol Metab* 47:675–80

Arato M, O'Connor R, Meltzer H, Bradbury J (1997a) Ziprasidone: efficacy in the prevention of relapse and in the long-term treatment of negative symptoms of chronic schizophrenia, *Eur Neuropsychopharmacol* 7(Suppl 2):S214

Arato M, O'Connor R, Bradbury JE, Meltzer H for the Zeus Study Group (1997b) Ziprasidone in the long term treatment of negative symptoms and prevention of exacerbation of schizophrenia, *Eur Neuropsychopharmacol* 7(Suppl 2):S214

Arnt J, Skarsfeldt T (1998) Do novel antipsychotics have similar pharmacological characteristics? A review of the evidence, *Neuropsychopharmacology* 18:63–101

Arvanitis LA, Miller BG and the Seroquel Trial 13 Study Group (1997) Multiple fixed doses of Seroquel (Quetiapine) in patients with acute exacerbation of schizophrenia: a comparison with haloperidol and placebo, *Biol Psychiatry* 42:233–46

Arvanitis LA, Rak IW (1998) Efficacy, safety and tolerability of 'Seroquel' (Quetiapine) in elderly subjects with psychotic disorder. 151st American Psychiatric Association Annual Meeting, Toronto, Canada, 4 June 1998

Ataya K, Mercado A, Kartaginer J et al (1988) Bone density and reproductive hormones in patients with neuroleptic-induced hyperprolactinemia, *Fertil Steril* 50:876–81.

Aweeka F, Horton M, Swan S et al (1997) The pharmacokinetics of ziprasidone in subjects with normal and impaired renal function, *Eur Neuropsychopharmacol* 7(Suppl 2):S214

Azmitia EC, Whitaker-Azmitia PM (1996) Anatomy, cell biology, and plasticity of the serotonergic system: neuropsychopharmacological implications for the actions of psychotropic drugs. In: Bloom FE, Kupfer DJ, eds, *Psychopharmacology: The Fourth Generation of Progress* (Raven Press: New York) 443–50

Baldessarini RJ (1995) Dopamine receptors and clinical medicine. In: Meade KA, Meade RL, eds, *Dopamine Receptors* (Humana Press: Totowa) 457–98

Baldessarini RJ (1996) Drugs and the treatment of psychiatric disorders: psychosis and anxiety. In: Hardman JG, Limbird LE, eds, *Goodman and Gilman's Pharmacological Basis of Therapeutics* 9th edn (McGraw Hill) 399–430.

Beasley Jr CM, Sanger T, Satterlee W et al and the Olanzapine HGAP Study Group (1996a) Olanzapine versus placebo: results of a double-blind fixed-dose olanzapine trial, *Psychopharmacology* 124:159-67

Beasley CM, Tollefson GD, Tran PV (1997) Efficacy of olanzapine: an overview of pivotal clinical trials, *J Clin Psychiatry* 58(Suppl 10):7–12

Beasley Jr CM, Tollefson G, Tran P et al and the Olanzapine HGAD Study Group (1996b) Olanzapine versus placebo and haloperidol: acute phase results of the North American double-blind olanzapine trial, *Neuropsychopharmacology* 14(2):111-23

Bench CJ, Lammertsma AA, Grasby PM et al (1996) The time course of binding to striatal dopamine D2 receptors by the neuroleptic ziprasidone (CP-88,059-01) determined by positron emission tomography, *Psychopharmacology* 124:141-7

Beuzen J-N, Birkett MA, Kiesler GM, Wood AJ (1998) Olanzapine vs clozapine: a double-blind international study in the treatment of resistant schizophrenic patients. ACNP Meeting, Las Croabas, Puerto Rico, 13-18 December

Biller BMK, Baum HBA, Rosenthal DI et al (1992) Progressive trabecular osteopenia women with hyperprolactinaemic amenorrhea, *J Clin Endocrinol Metab* 75:692-7.

Bishop MP, Simpson GM, Dunnett CW, Kiltie H (1977) Efficacy of loxapine in the treatment of paranoid schizophrenia, *Psychopharmacology* 51:107-15

Bitello B, Martin A, Hill J et al (1997) Cognitve and behavioral effects of cholinergic dopaminergic and serotonergic blockade in humans, *Neuropsychopharmacology* 16:15-24

Bloom FE, Kupfer D, eds (1996) *Psychopharmacology: The Fourth Generation of Progress* (Raven Press: New York)

Bond AJ, Lader MH (1996) *Understanding Drug Treatment and Mental Health Care* (John Wiley and Sons: Chichester)

Bondolfi G, Dufour H, Patris M et al on behalf of the Risperidone Study Group (1998) Risperidone versus clozapine in treatment-resistant chronic schizophrenia: a randomized double-blind study, *Am J Psychiatry* 155(4):499-504

Borison, RL, Arvanitis LA, Miller BG and the US Seroquel Study Group (1996) ICI 205,636, an atypical antipsychotic: efficacy and safety in a multicenter, placebo-controlled trial in patients with schizophrenia, *J Clin Psychopharmacol* 16(2):158-69

Bouloux PM, Grossman A (1987) Hyperprolactinaemia and sexual function in the male, *Br J Hosp Med* 37(6):503-10

Branchey MH, Lee JH, Simpson GM et al (1978) Loxapine succinate as a neuroleptic agent: evaluation in two populations of elderly psychiatric patients, *J Am Geriatr Soc* 26(6):263-7

Brook S, Krams M, Gunn KP and the Ziprasidone IM Study Group (1998) The efficacy and tolerability of intramuscular (IM) ziprasidone versus IM haloperidol in patients with acute, non-organic psychosis, *Eur Psychiatry* 13(4):303s

Buckley PF (1998) Substance abuse in schizophrenia: a review, *J Clin Psychiatry* 59(Suppl 3):26-30

Buckley PF (in press) The role of typical and atypical antipsychotic medications in the management of agitation and aggression, *J Clin Psychiatry*

Byerly MJ, DeVane L (1996) Pharmacokinetics of clozapine and risperidone: a review of recent literature, *J Clin Psychopharmacol* 16(2):177-87

Bymaster FP, Calligaro DO, Falcone JF et al (1996) Radioreceptor binding profile of the atypical antipsychotic olanzapine, *Neuropsychopharmacology* 14(2):87-96

Carlyle W, Ancill RJ, Sheldon L (1993) Aggression in the demented patient: a double-blind study of loxapine versus haloperidol, *Int Clin Psychopharmacol* 8:103-8

Casey DE (1998) Effects of clozapine therapy in schizophrenic individuals at risk for tardive dyskinesia, *J Clin Psychiatry* 59(Suppl 3):31-7

Chappell P, Sallee F (1998) The tolerability and efficacy of ziprasidone in the treatment of children and adolescents with Tourette's syndrome (TS), *Eur Psychiatry* 13(Suppl 4):313

Charalampous KD, Freemeser GF, Malev J,

Ford K (1974) Loxapine succinate: a controlled double-blind study in schizophrenia, *Curr Ther Res* 16(8):829-37

Cheung SW, Tang SW, Remington G (1991) Simultaneous quantitation of loxapine, amoxapine and their 7- and 8-hydroxy metabolites in plasma by high-performance liquid chromatography, *J Chromatogr* 564:213-21

Chouinard G, Jones BJ, Remington G et al (1993) A Canadian multicenter placebo-controlled study of fixed doses of risperidone and haloperidol in the treatment of chronic schizophrenic patients, *J Clin Psychopharmacol* 13(1):25-40

Chung Y-C, Eun H-B (1998) Hyperprolactinaemia induced by risperidone, *Int J Neuropsychopharm* 1:93-4

Civelli O (1996) Molecular biology of the dopamine receptor subtypes. In: Bloom FE, Kupfer J, eds, *Psychopharmacology: The Fourth Generation of Progress* (Raven Press: New York) 155-62

Coccaro EF (1998) Clinical outcome of psychopharmacologic treatment of borderline and schizotypal personality disordered subjects, *J Clin Psychiatry* 59(Suppl 1):30-37

Cohen BM, Harris PQ, Altesman RI, Cole JO (1982) Amoxapine: neuroleptic as well as antidepressant?, *Am J Psychiatry* 139(9):1165-7

Cole JO, Swett C, Campbell C, Trenholm MS (1982) Parenteral and oral loxitane in the treatment of schizophrenic disorders, *Curr Ther Res* 31(4):656-61

Conley RR (1998) Optimizing treatment with clozapine, *J Clin Psychiatry* 59(Suppl 3):44-8

Conley RR, Brecher M and the Risperidone/Olanzapine Study Group (1998a) Risperidone versus olanzapine in patients with schizophrenia or schizoaffective disorder. 11th ECNP Congress, Paris, France, 31 October–4 November 1998

Conley RR, Tamminga CA, Bartko JJ et al (1998b) Olanzapine compared with chlorpromazine in treatment resistant schizophrenia, *Am J Psychiatry* 155(7):914–20

Cooper JR, Bloom FE, Roth RH (1996) *The Biochemical Basis of Neuropharmacology* 7th edn (Oxford University Press: New York)

Coryell W (1998) The treatment of psychotic depression, *J Clin Psychiatry* 59(Suppl 1):22–9

Coupet J, Rauh CE (1979) ³H-Spiroperidol binding to dopamine receptors in rat striatal membranes: influence of loxapine and its hydroxylated metabolites, *Eur J Pharmacol* 55:215–18

Coupet J, Rauh CE, Szues-Myers VA ,Yunger LM (1979) 2-Chloro-11-(1–piperazinyl) dibenz {b,f} {1,4} loxazepine (Amoxapine), an antidepressant with antipsychotic properties – a possible role for 7-hydroxyamoxapine, *Biochem Pharmacol* 28:2514–15

Coupet J, Szues-Myers VA, Greenblatt EN (1976) The effects of 2–chloro-11–L(4-methyl-l-piperazinyl)-dibenz {b,f} {1,4} loxazepine (loxapine) and its derivatives on the dopamine-sensitive adenylate cyclase of rat striatal homogenates, *Brain Res* 116:177–80

Crawford AM, Beasley CM, Tollefson GD (1997) The acute and long term effects of olanzapine compared with placebo and haloperidol on serum prolactin concentrations, *Schizophr Res* 26:41–54

Dahl SG (1982) Active metabolites of neuroleptic drugs: possible contribution to therapeutic and toxic effects, *Ther Drug Monit* 4:33–40

Daniel D, Reeves K, Harrigan EP (1997) The efficacy and safety of ziprasidone

80 mg/day and 160 mg/day in schizophrenia and schizoaffective disorder, *Schizophr Res* 24(1,2):77

Daniel DG, Reeves KR, Swift RH, Harrigan EP (1998) Rapid-acting intramuscular ziprasidone: an overview. ACNP Meeting, Las Croabas, Puerto Rico, 13–18 December 1998

Daniel DG, Whitcomb SR (1998) Treatment of the refractory schizophrenic patient, *J Clin Psychiatry* 59(Suppl 1):13–21

Daniel DG, Zimbroff DL, Potkin SG et al and the Ziprasidone Study Group (in press) Ziprasidone 80 mg/day and 160 mg/day in the acute exacerbation of schizophrenia and schizoaffective disorder: a 6 week placebo-controlled trial, *Neuropsychopharmacology*.

Dean GA, Gallant DM (1979) Intramuscular loxapine (Loxitane): rapid tranquilization of acutely disturbed schizophrenic patients, *Curr Ther Res* 25(6):721–5

Degen K (1982) Sexual dysfunction in women using major tranquilizers, *Psychosomatics* 23:959–61

Dellenberg AJ, Hopkins HS (1996) Antipsychotics and bipolar disorder, *J Clin Psychiatry* 57(Suppl 9):49–52, 62

Deniker P, Loo H, Cottereau MJ (1980) Parenteral loxapine in severely disturbed schizophrenic patients, *J Clin Psychiatry* 41(1):23–6

DePaulo Jr JR, Ayd Jr FJ (1982) Loxapine: fifteen years' clinical experience, *Psychosomatics* 23(3):261–71

Deutsch AY, Lee MC, Iadarola MJ (1992) Regionally specific effects of atypical antipsychotic drugs on striatal FOS expression: the nucleus accumbens shell is a locus of antipsychotic action, *Mol Cell Neurosci* 3:332–41

Doss FW (1979) The effect of antipsychotic drugs on body weight: a retrospective review, *J Clin Psychiatry* 40:528–30

Dubin WR, Weiss KJ (1986) Rapid tranquilization: a comparison of thiothixene with loxapine, *J Clin Psychiatry* 47(6):294-7

Ereshefsky L, Lacombe S (1993) Pharmacological profile of risperidone, *Can J Psychiatry* 38(Suppl 3):S80-S88

Ereshefsky L (in press) Pharmacologic and pharmacokinetic considerations in choosing an antipsychotic. *J Clin Psychiatry.*

Everson G, Lasseter KC, Anderson KE et al (1997) The pharmacokinetics of ziprasidone in subjects with normal and impaired hepatic function, *Eur Neuropsychopharmacol* 7(Suppl 2):S220

Farde L, Nyberg S, Oxenstierna G et al (1995) Positron emission tomography studies on D_2 and 5-HT$_2$ receptor binding in risperidone-treated schizophrenic patients, *J Clin Psychopharmacol* 15(Suppl 1):19S-23S

Filho UV, Caldeira VV, Bueno JR (1975) The efficacy and safety of loxapine succinate in the treatment of schizophrenia: a comparative study with thiothixene, *Curr Ther Res* 18(3):476-90

Findling RL, Grcevich SJ, Lopez I, Schulz SC (1996) Antipsychotic medications in children and adolescents, *J Clin Psychiatry* 57(Suppl 9):19-23, 60

Fischman AJ, Bonab AA, Babich JW et al (1996) Positron emission tomographic analysis of central 5-hydroxytryptamine 2 receptor occupancy in healthy volunteers treated with the novel antipsychotic agent ziprasidone, *J Pharmacol Exp Ther* 279(2):939-47

Fritze J, Bandelow B (1998) The QT interval and the atypical antipsychotic, sertindole, *Int J Psychiatry Clin Pract* 2:265-73

Fruensgaard K, Jensen K (1976) Treatment of acute psychotic patients with loxapine parenterally, *Curr Ther Res* 19(2):164-9

Fruensgaard K, Korsgaard S, Jorgensen H, Jensen K (1977) Loxapine versus haloperidol parenterally in acute psychosis with agitation, *Acta Psychiatria Scand* 56(4):256-64

Fulton A, Norman T, Burrows GD (1982) Ligand binding and platelet uptake studies of loxapine, amoxapine and their 8 hydroxylated derivatives, *J Affective Disord* 4:113-19

Gefvert O, Berstrom M, Langstrom B et al (1998) Time course of central nervous dopamine-D_2 and 5-HT$_2$ receptor blockade and plasma drug concentrations after discontinuation of quetiapine(Seroquel®) in patients with schizophrenia, *Psychopharmacology* 135:119-26

Gelenberg AJ, Bassuk EL (1997) *The Practitioner's Guide to Psychoactive Drugs* 4th edn (Plenem Press: New York)

Ghadirian AM, Choiunard G, Annable L (1982) Sexual dysfunction and plasma prolactin levels in neuroleptic-treated schizophrenic outpatients, *J Nerv Ment Dis* 170:463-7

Gheuens J, Grebb JA (1998) Comments on article by Tran and associates 'Double-blind comparison of olanzapine versus risperidone in treatment of schizophrenia and other psychotic disorders', *J Clin Psychopharmacol* 18(2):176-7

Glazer WM (1997) Olanzapine and the new generation of antipsychotic agents: pattterns of use, *J Clin Psychiatry* 58(Suppl 10):18-21

Glazer WM (in press) Does loxapine have 'atypical' properties? Clinical evidence, *J Clin Psychopharmacol*

Glazer WM, Dickson RA (1998) Clozapine reduces violence and persistent aggression in schizophrenia, *J Clin Psychiatry* 59(Suppl 3):8-14

Glennon RA, Dukat M (1996) Serotonin

receptor subtypes. In: Bloom FE, Kupfer DJ, eds, *Psychopharmacology: The Fourth Generation of Progress* (Raven Press: New York) 415–30

Goff DC, Posover T, Herz L et al (1998) An exploratory, haloperidol-controlled, dose-finding study of ziprasidone in hospitalized patients with schizophrenia or schizoaffective disorder, *J Clin Psychopharmacol* 18(4):296–304

Goldberg TE, Weinberger DR (1996) Effects of neuroleptic medications on the cognition of patient with schizophrenia: a review of recent studies, *J Clin Psychiatry* 57(Suppl 9):62–5

Goldstein JM (1998a) 'Seroquel' (quetiapine fumarate) reduces hostility and aggression in patients with acute schizophrenia. American Psychiatric Association 151st Annual Meeting, Toronto, Canada, 4 June 1998

Goldstein JM (1998b) Quetiapine fumarate: effects of hostility, aggression and affective symptoms in patients with acute schizophrenia. NCDEU 38th Annual Meeting, Boca Raton, Florida, 10–15 June 1998

Goldstein JM (1998c) Low incidence of reproductive/hormonal side effects with Seroquel (quetiapine fermonate) is supported by its lack of elevation of plasma prolactin concentrations. American Psychiatric Association 151st Annual Meeting, Toronto, Canada, 4 June 1998

Grace AA, Bunney BS (1996) Electrophysiological properties of mid-brain dopamine neurons. In: Bloom FE, Kupfer J, eds, *Psychopharmacology: The Fourth Generation of Progress* (Raven Press: New York) 163–77

Green MF, Marshall J. BD, Wirshing WC et al (1997) Does risperidone improve verbal working memory in treatment-resistant schizophrenia? *Am J Psychiatry* 154(6):799–804

Greenspan SL, Neer RM, Ridgway EC,

Klibanski A (1986) Osteoporosis in men with hyperprolactinemic hypogonadism, *Ann Intern Med* 104:777–82.

Greenspan SL, Oppenheim DS, Klibanski A (1989) Importance of gonadal steroids to bone mass in men with hyperprolactinemic hypogonadism, *Ann Intern Med* 110:526–31.

Grohmann, R, Schmidt LG, Spieb-Kiefer C, Ruther E (1989) Agranulocytosis and significant leukopenia with neuroleptic drugs: results from the AMUP program, *Psychopharmacology* 99:S109–S112

Gunn KP, Zorn SH, Heym J (1997) Ziprasidone: preclinical profile of a new antipsychotic agent, *Schizophr Res* 24(1,2):204

Halbreich U, Rojansky N, Palter S et al (1995) Decreased bone mineral density in medicated psychiatric patients, *Psychosom Med* 57:485–91

Halbreich U, Palter (1996) Accelerated osteoporosis in psychiatric patients: possible pathophysiological processes, *Schizoph Bull* 22:447–54.

Hale AS (1998) A review of the safety and tolerability of sertindole, *Int J Clin Psychopharmacol* 13(Suppl 3):S65–S70

Hamilton SH, Revicki DA, Genduso LA, Beasley Jr CM (1998) Olanzapine versus placebo and haloperidol: quality of life and efficacy results of the North American double-blind trial, *Neuropsychopharmacology* 18(1):41–9

Harvey PD, Keefe RS (1997) Cognitive impairment in schizophrenia and implications of atypical neuroleptic treatment, *CNS Spectrums* 2:41–55

Hillard JR (1998) Emergency treatment of acute psychosis, *J Clin Psychiatry* 59(Suppl 1):57–61

Honigfeld G, Arellano F, Sethi J et al (1998) Reducing clozapine-related morbidity and mortality: 5 years of experience with the Clozaril National Registry, *J Clin Psychiatry* 59(Suppl 3):3–7

Hue B, Palomba B, Giacardy-Paty M (1998) Concurrent high-performance liquid chromatographic measurement of loxapine and amoxapine and of their hydroxylated metabolites in plasma, *Ther Drug Monit* 20(3):335–9

Hyman SE, Arana GW, Rosenbaum JF (1995) *Handbook of Psychiatric Drug Therapy* 3rd edn (Little Brown: Boston)

Jacobs BL, Fornal CA (1996) Serotonin and behavior: a general hypothesis. In: Bloom FE, Kupfer DJ, eds, *Psychopharmacology: The Fourth Generation of Progress* (Raven Press: New York) 461–70

Janicak PG, Davis JM, Preskorn SH, Ayd FJ (1997) *Principles and Practice of Psychopharmacotherapy* (Williams and Wilkins Press: Baltimore, MD)

Kahn RS, Davis KL (1996) New developments in dopamine and schizophrenia. In: Bloom FE, Kupfer J, eds, *Psychopharmacology: The Fourth Generation of Progress* (Raven Press: New York) 1193–1203

Kane JM (1997) Sertindole: a review of clinical efficacy, *Int J Clin Psychopharmacol* 13(Suppl 3):S59–S64

Kane J, Honigfeld G, Singer J, Meltzer HY and the Clozaril Collaborative Study Group (1988) Clozapine for the treatment-resistant schizophrenic. A double-blind comparison with chlorpromazine, *Arch Gen Psychiatry* 45:789–96

Kapur S (1996) Serotonin 2 antagonism and EPS benefits: is there a causal connection?, *Psychopharmacology* 124:35–9

Kapur S, Remington G (1996) Serotonin dopamine interaction and its relevance to schizophrenia, *Am J Psychiatry* 153:466–76

Kapur S, Zipoursky RB, Jones C et al (1996) The D_2 receptor occupancy profile of loxapine determined using PET, *Neuropsychopharmacology* 15(6):562–6

Kapur S, Zipurski R, Remington G et al (1997) PET evidence that loxapine is an equipotent blocker of 5HT2 and D2 receptors: implications for the therapeutics of schizophrenia, *Am J Psychiatry* 154:1525–9

Kasper S, Kufferle B (1998) Comments on 'Double-blind comparison of olanzapine versus risperidone in the treatment of schizophrenia and other psychotic disorders' by Tran and associates, *J Clin Psychopharmacol* 18(4):353–4

Kasper S, Tauscher J, Kufferle B et al (1998) Sertindole and dopamine D_2 receptor occupancy in comparison to risperidone, clozapine and haloperidol— a ^{123}I-IBZM SPECT study, *Psychopharmacology* 136:367–73

Keck PE, McElroy SL, Stakowski SM (1996) New developments in the pharmacological treatment of schizo affective disorder, *J Clin Psychiatry* 57(Suppl 9):41–8

Keck PE, McElroy SL, Strakowski SM, West SA (1994) Pharmacological treatment of schizo affective disorder, *Psychopharmacology* 114:529–38

Keck PE, Reeves KR, Harrigan EP, Daniel D (1998) The efficacy of ziprasidone in the treatment of patients with an acute exacerbation of schizophrenia or schizoaffective disorder. APA Institute for Psychiatric Services, Los Angeles, California, 2–4 October 1998

Kee KS, Kern RS, Marshall Jr BD, Green MF (1998) ʟ Risperidone versus haloperidol for perception of emotion in treatment-resistant schizophrenia: preliminary findings, *Schizophr Res* 31:159–65

Kelly DL, Conley RR, Love RC, Bartko JJ (1998) Dose–outcome analysis of risperidone. 37th Annual Meeting of the American College of Neuropsychopharmacology, Las Croabas, Puerto Rico, 14-18 December, 1998

Kiloh LG, Williams SE, Grant DA, Whetton PS (1976) A double-blind comparative trial of loxapine and trifluoperazine in acute and chronic schizophrenic patients, *J Int Med Res* 4(6)441-8

Kilpatrick IC, Rowley HL, Needham PL, Heal DJ (1998a) Zotepine enhances noradrenaline levels in rat frontal cortex microdialysates: further support for antidepressant activity. 11th Annual ECNP Congress, Paris, France 31 October-4 November 1998

Kilpatrick AT, Welch CP, Butler A, Tweed JA (1998b) The efficacy of zotepine in reducing BPRS total scores. 11th Annual ECNP Congress, Paris, France, 31 October-4 November 1998

King DJ, Link CGG, Kowalcyk B (1998) A comparison of bd and tid dose regimens of quetiapine (Seroquel®) in the treatment of schizophrenia, *Psychopharmacology* 137:139-46

Kinon BJ, Basson B, Tollefson GD (1998) Gender-specific prolactin response to treatment with olanzapine versus haloperidol in schizophrenia. 151st American Psychiatric Association Meeting, Toronto, Canada, 30 May-4 June 1998

Kinon BJ, Lieberman JA (1996) Mechanisms of action of atypical antipsychotic drugs: a critical analysis, *Psychopharmacology* 124:2-34

Kleinberg DL, Brecher M, Davis JM (1997) Prolactin levels and adverse events in patients treated with risperidone. 150th Annual Meeting of the American Psychiatric Association, San Diego, California, 17-22 May 1997

Klibanski A, Biller BMK, Rosenthal DI et al (1988) Effects of prolactin and estrogen deficiency in amenorrheic bone loss, *J Clin Endocrinol Metab* 67: 124-30

Klibanski A, Greenspan SL (1986) Increase in bone mass after treatment of hyperprolactinemic amenorrhea, *N Engl J Med* 315:542-6

Klibanski A, Neer RM, Beitins IZ et al (1980) Decreased bone density in hyperprolactinaemic women, *N Engl J Med* 303:1511-14

Knable MB, Kleinman JE, Weinberger DR (1998) Neurobiology of schizophrenia. In: Schatzberg AF, Nemeroff CB, eds, *Textbook of Psychopharmacology* 2nd edn (American Psychiatric Press: Washington DC) 589-608

Koppelman MCS, Kurtz DW, Morrish KA et al (1984) Vertebral body bone mineral content in hyperprolactinemic women, *J Clin Endocrinol Metab* 59:1050-3.

Kramer M, Roth T, Salis PJ, Zorick FJ (1978) Relative efficacy and safety of loxapine succinate (Loxitane) and thioridazine hydrochloride (Mellaril) in the treatment of acute schizophrenia, *Curr Ther Res* 23(5):619-31

Kufferle B, Tauscher J, Asenbaum S et al (1997) IBZM SPEC imaging of striatal dopamine 2 receptors in psychotic patients treated with a novel antipsychotic substance quetiapine in comparison to clozapine and haloperidol, *Psychopharmacology* 133:323-8

Lehmann CR, Ereshefsky L, Saklad SR, Mings TE (1981) Very high dose loxapine in refractory schizophrenic patients, *Am J Psychiatry* 138(9):1212-14

Lehmann HE (1986) Neuroleptics and sexual functioning, *Integ Psychiatry* 4:96-108

LeMoal M (1996) Mesocorticolimbic dopaminergic neurons: functional and regulatory roles. In: Bloom FE, Kupfer J, eds, *Psychopharmacology: The Fourth Generation of Progress* (Raven Press: New York) 283-94

Leonard BE (1997) *Fundamentals of*

Psychopharmacology 2nd edn (John Wiley & Sons: Chichester)

Leonard MP, Nickel CJ, Morales A (1989) Hyperprolactinaemia and impotence: why, when and how to investigate, *J Urol* 142:992-4

Leone NF (1979) Open evaluation of loxapine succinate (Loxitane) in the treatment of acutely ill schizophrenic outpatients, *Curr Ther Res* 26(5):515-24

Lewis DA (1997) Development of the prefrontal cortex during adolescence: insights into vulnerable neurocircuits in schizophrenia, *Neuropsychopharmacology* 16:385-98

Leysen JE, Janssen PMF, Heylen L et al (1998) Receptor interactions of new antipsychotics: relation to pharmacodynamic and clinical effects, *Int J Psychiatry Clin Pract* 2(Suppl 1):S3-S18

Leysen JE, Janssen PMF, Megens AHHP, Schotte A (1994) Risperidone: a novel antipsychotic with balanced serotonin dopamine antagonism, receptor occupancy profile, and pharmacological activity, *J Clin Psychiatry* 55(Suppl 5):5-12

Li X-M, Perry KW, Wong DT, Bymaster FP (1998) Olanzapine increases in vivo dopamine and norepinephrine release in rat prefrontal cortex, nucleus accumbens and striatum, *Psychopharmacology* 136:153-61

Lieberman JA (1998) Maximizing clozapine therapy: managing side effects, *J Clin Psychiatry* 59(Suppl 3):38-43

Lieberman JA, Kane JM, Johns CA (1989) Clozapine: guidelines for clinical management, *J Clin Psychiatry* 50(9):329-38

Lieberman JA, Sheitman B, Chakos M (1998) The development of treatment resistance in patients with schizophrenia: a clinical and pathophysiologic perspective, *J Clin Psychopharmacol* 18(2)(Suppl 1):20S-24S

Lieberman JA, Sheitman BB, Kinon BJ (1997) Neurochemical sensitization in the pathophysiology of schizophrenia: deficits in dysfunction in neuronal regulation and plasticity, *Neuropsychopharmacology* 17:205-29

Lindenmayer JP, Iskander A, Park M et al (1998) Clinical and neurocognitive effects of clozapine and risperidone in treatment-refractory schizophrenic patients: a prospective study, *J Clin Psychiatry* 59(10):521-7

Littrell KH, Johnson CG, Littrell S, Peabody CD (1998) Marked reduction of tardive dyskinesia with olanzapine, *Arch Gen Psychiatry* 55(3):279-80

Lussier I, Stip E (1998) The effect of risperidone on cognitive and psychopathological manifestations of schizophrenia, *CNS Spectrums* 3(10):55-69

McConville B, Arvantis L, Wong J et al (1998) Pharmacokinetics, tolerability and clinical effectiveness of quetiapine fumarate in adolescents with selected psychotic disorders. American Psychiatric Association 151st Annual Meeting, Toronto, Canada, 4 June 1998

Macgibbon GA, Lawlor PA, Bravo R, Dragunow M (1994) Clozapine and haloperidol produce a differential pattern of immediate early gene expression in rat caudate putamen, nucleus accumbens, lateral septum, and islands of calleja, *Mol Brain Res* 23:21-32

Mansour A, Meador-Woodruff JH, Lopez JF, Watson SJ (1998) Biochemical anatomy: insights into the cell biology and pharmacology of the dopamine and serotonin systems in the brain. In: Schatzberg AF, Nemeroff CB, eds, *Textbook of Psychopharmacology* 2nd edn (American Psychiatric Press: Washington DC) 55-74

Mansour A, Watson SJ (1996) Dopamine receptor expression in the central nervous system. In: Bloom FE, Kupfer J, eds, *Psychopharmacology: The Fourth Generation of Progress* (Raven Press: New York) 207-19

Marcus MM, Nomikos GG, Svensson TH (1996) Differential actions of typical and atypical antipsychotic drugs on dopamine release in the core and shell of the nucleus accumbens, *Eur Neuropsychopharmacol* 6:29-38

Marder SR (1998a) Antipsychotic medications. In: Schatzberg AF, Nemeroff CB, eds, *Textbook of Psychopharmacology* 2nd edn (American Psychiatric Press: Washington DC) 309-22

Marder SR (1998b) Facilitating compliance with antipsychotic medication, *J Clin Psychiatry* 59(Suppl 3):21-5

Marder SR, Davis JM, Chouinard G (1997) The effects of risperidone on the 5 dimensions of schizophrenia derived by factor analysis: combined results of the North American trials, *J Clin Psychiatry* 58:538-46.

Marder SR, Meibach RC (1994) Risperidone in the treatment of schizophrenia, *Am J Psychiatry* 151(6):825-35

Marken PA, Haykal RF, Fisher JN (1992) Management of psychotropic-induced hyperprolactinaemia, *Clin Pharmacol Ther* 11:851-6.

Martin J, Gomez JC, Garcia-Bernardo E et al and the Spanish Group for the Study of Olanzapine in Treatment Refractory Schizophrenia (1997) *J Clin Psychiatry* 58(11):479-83

Masand PS (1998) Weight gain associated with atypial antipsychotics, *J Psychot Disord* II(3):4-6

Meats P (1997) Quetiapine ('Seroquel'); an effective and well tolerated atypical antipsychotic, *Int J Psychiatry Clin Pract* 1:231-9

Meltzer HY (1991) Dopaminergic and serotonergic mechanisms and the action of clozapine. In: Tamminga CA, Schulz SC, eds, *Advances in Neuropsychiatry and Psychopharmacology* Vol 1 (Raven Press: New York) 333-40

Meltzer HY (1994) An overview of the mechanism of action of clozapine, *J Clin Psychiatry* 55(9)(Suppl B):47-52

Meltzer HY (1996) Atypical antipsychotic drugs. In: Bloom FE, Kupfer J, eds, *Psychopharmacology: The Fourth Generation of Progress* (Raven Press: New York) 1277-86

Meltzer HY (1998) Suicide in schizophrenia: risk factors and clozapine treatment, *J Clin Psychiatry* 59(Suppl 3):15-20

Meltzer HY, Jayathilake K (in press) Low dose loxapine in the treatment of schizophrenia: is it effective and more 'atypical' than standard dose loxapine? *J Clin Psychopharmacol*

Meltzer HY, Lee M, Cola P (1998) The evolution of treatment resistance: biological implications, *J Clin Psychopharmacol* 18(2):Suppl 1):2S-4S

Meltzer HY, Matsubari F, Lee JC (1989) The ratios of serotonin 2 and dopamine 2 affinities differentiate atypical and typical antipsychotic drugs, *Psychopharmacol Bull* 25:390-92

Meltzer HY, Stahl SM (1976) The dopamine hypothesis of schizophrenia: a review, *Schizophr Bull* 2(1):19-76

Miceli JJ, Gunn KP, Rubin RH et al (1997) 5HT2 and D2 receptor occupancy of ziprasidone in healthy volunteers, *Schizophr Res* 24(1,2):178

Midha KK, Hubbard JW, McKay G et al (1993) The role of metabolites in a bioequivalence study 1: loxapine, 7-hydroxyloxapine and 8-hydroxyloxapine, *Int J Clin Pharmacol, Ther Toxicol* 31(4):177-83

Mitchell JE, Popkin MK (1982) Antipsychotic drug therapy and sexual dysfunction in men, *Am J Psychiatry* 139:633-7

Moghaddam B, Bunney BS (1990) Acute effects of typical and atypical antipsychotic drugs on the release of dopamine from prefrontal cortex, nucleus accumbens and striatum of the rat: and in vitro microdialysis study, *J Neurochem* 54:1755-60

Mowerman S, Siris SG (1996) Adjunctive loxapine in a clozapine-resistant cohort of schizophrenic patients, *Ann Clin Psychiatry* 8(4):193-7

Moyano CZ (1975) A double-blind comparison of loxitane loxapine succinate and trifluoperazine hydrochloride in chronic schizophrenic patients, *Dis Nerv Syst* 36(6):301-4

Muirhead GJ, Hold PR, Oliver S et al (1996) The effect of ziprasidone on steady state pharmacokinetics of a combined oral contraceptive, *Eur Neuropsychopharmacol* 6(Suppl 3):38

Murray RM, Van Os J (1998) Predictors of outcome in schizophrenia, *J Clin Psychopharmacol* 18(2)(Suppl 1):2S-4S

Needham PL, Skill MJ, Heal DJ (1998) Zotepine: preclinical tests predict antipsychotic efficacy and an atypical profile. 11th Annual ECNP Congress, Paris, France, 31 October-4 November 1998

Nelson MW, Kelley DL, Love RC, Conley RR (1998) Risperidone versus olanzapine: discharge rates and economic considerations. 37th Annual Meeting of the American College of Neuropsychopharmacology, Las Croabas, Puerto Rico, 14-18 December 1998

Nestoros JN, Lehmann HE, Ban TA (1981) Sexual behavior of the male schizophrenic: the impact of illness and medications, *Arch Sex Behav* 10:421-42

Netto NR, Claro JA (1993) The importance of hyperprolactinaemia in impotence, *Rev Paul Med* 111:454-5

Nguyen TY, Kosofsky BE, Birnbaum R et al (1992) Differential expression of CFOS and Zif 268 in rat striatum, after haloperidol, clozapine and amphetamine, *Proc Nat Acad Sci USA* 89:4270-74

Nordstrom AL, Farde L, Halldin C (1993) High 5-HT^2 receptor occupancy in clozapine treated patients demonstrated by PET, *Psychopharmacology* 110:365-7

Nyberg S, Farde L, Eriksson L et al (1993) 5HT2 and D2 dopamine receptor occupancy in the living human brain: a PET study with risperidone, *Psychopharmacology* 110:265-72

Nyberg S, Farde L, Halldin C (1997) A PET study of 5HT2 and D2 dopamine receptor occupancy induced by olanzapine in healthy subjects, *Neuropsychopharmacology* 16:1-7

Nystrom E, Leman J, Lindquist O et al (1988) Bone mineral content in normally menstruating women with hyperprolactinaemia, *Horm Res* 29:214-17.

O'Connell RA, Lieberman JA (1978) Parenteral loxapine in acute schizophrenia, *Curr Ther Res* 23(2):236-42

Owens DGC (1994) Extrapyramidal side effects and tolerability of risperidone: a review, *J Clin Psychiatry* 55(Suppl 5):29-35

Owens MJ, Risch SD (1998) Atypical antipsychotics. In: Schatzberg AF, Nemeroff CB, eds, *Textbook of Psychopharmacology* 2nd edn (American Psychiatric Press: Washington DC) 323-48

Palmgren K, Tweed JA, Welch CP et al (1998) A multicentre naturalistic long term study of zotepine. 11th Annual ECNP Congress; Paris, France, 31 October-4 November 1998

Paprocki J, Barcala Peixooto MP, Mendes Andrade N (1976) A controlled double-blind comparison between loxapine and haloperidol in acute newly hospitalized schizophrenic patients, *Psychopharmacol Bull* 12(2):32-4

Paprocki J, Versiani M (1977) A double blind comparison between loxapine and haloperidol by parenteral route in acute schizophrenia, *Curr Ther Res* 21(1):80–100

Petit M, Raniwalla J, Tweed J et al (1998) A comparison of an atypical and typical antipsychotic, zotepine versus haloperidol, in patients with acute exacerbation of schizophrenia. 11th Annual ECNP Congress, Paris, France, 31 October–4 November 1998

Peuskens J (1995) Risperidone in the treatment of patients with chronic schizophrenia: a multi-national, multi-centre, double-blind, parallel-group study versus haloperidol, *Br J Psychiatry* 166:712–26

Peuskens J, Link CGG (1997) A comparison of quetiapine and chlorpromazine in the treatment of schizophrenia, *Acta Psychiatrica Scand* 96(4):265–73

Pilowsky LS, Busatto GF, Taylor M et al (1996) Dopamine D$_2$ receptor occupancy in vivo by the novel atypical antipsychotic olazapine: a 123i IBZM single photon emission tomography study, *Psychopharmacology* 124:148–53

Pilowsky LS, O'Connell P, Davies N et al (1997) In vivo effects of striatal dopamine D$_2$ receptor binding by the novel atypical antipsychotic drug sertindole—a ^{123}I IBZM single photon emission tomography (SPET) study, *Psychopharmacology* 130:152–8

Pohen M, Zarate CA (1998) Antipsychotic agents in bipolar disorder, *J Clin Psychiatry* 59(Suppl 1):38–49

Prakash C, Kamel A, Cui D et al (1997a) Ziprasidone metabolism and cytochrome P450 isoforms, *Biol Psychiatry* 42:405

Prakash C, Kamel A, Gummerus J, Wilner K (1997b) Metabolism and excretion of a new antipsychotic drug, ziprasidone, in humans, *Drug Metab Dispos* 25(7):863–72

Prakash A, Lamb HM (1998) Zotepine: a review of its pharmacodynamic and pharmacokinetic properties and therapeutic efficacy in the management of schizophrenia, *CNS Drugs* 9(2):153–75

Purdon SE (1998) Olanzapine vs risperidone vs haloperidol in early illness schizophrenia. 151st Annual American Psychiatric Meeting, Toronto, Canada

Quitken SM, Adams DC, Bowden CL et al (1998) *Current Psychotherapeutic Drugs* 2nd edn (American Psychiatric Press: Washington DC)

Reeves KR, Swift RH, Harrigan EP, Lesem M (1998a) A randomized, double-blind comparison of rapid acting, intramuscular ziprasidone 2 mg and 20 mg in patients with psychosis and acute agitation, *Eur Psychiatry* 13(4):303s–304s

Reeves KR, Swift RH, Harrigan EP (1998b) Intramuscular ziprasidone 10 mg and 20 mg in patients with psychosis and acute agitation. Presented at the 151st Annual Meeting of the American Psychiatric Association, 30 May–4 June 1998

Reeves KR, Swift RH and Harrigan EP (1998c) A comparison of rapid-acting, intramuscular (IM) ziprasidone 2 mg and 20 mg in patients with psychosis and acute agitation. Presented at the 151st Annual Meeting of the American Psychiatric Association, 30 May–4 June 1998

Reeves KR, Swift RH, Harrigan EP (1998d) Rapid acting intramuscular ziprasidone 10 mg and 20 mg in patients with psychosis and acute agitation, *Eur Psychiatry* 13(4):304s

Remington G, Kapur S (in press) Receptor effects of antipsychotics: bridging basic and clinical findings using PET, *J Clin Psychiatry*.

Reynolds CW, Kilpatrick AT, Bratty JR

(1998) Zotepine, a broad spectrum antipsychotic with low EPS. 11th Annual ECNP Congress, Paris, France, 31 October–4 November 1998

Ribot C, Tremollieres F, Pouilles JM (1994) The effect of obesity on postmenopausal bone loss and the risk of osteoporosis, *Adv Nutr Res* 9:257–71.

Richelson E (1996) Pre-clinical pharamacology of neuroleptics: focus on new generation compounds, *J Clin Psychiatry* 57:(Suppl 11):4–11

Richelson E (in press) Receptor pharmacology of neuroleptics: relation to clinical effects, *J Clin Psychopharmacol*

Ring BJ, Binkley SN, Vandenbranden M, Wrighton SA (1996) In vitro interaction of the antipsychotic agent olanzapine with human cytochromes P450 CYP2C19, CYP2D6 and CYP3A, *Br J Clin Pharmacol* 41:181–6

Robertson AG, Berry R, Meltzer HY (1982) Prolactin stimulating effects of amoxapine and loxapine in psychiatric patients, *Psychopharmacology* 78:287–92

Robertson GS, Fibiger HC (1992) Neuroleptics increase CFOS expression in the full brain: contrasting effects of haloperidol and clozapine, *Neuroscience* 46:325–8

Robertson GS, Fibiger HC (1996) Effects of olanzapine on regional CFOS expression in the rat forebrain, *Neuropsychopharmacology* 14:105–10

Robertson GS, Matsubura H, Fibiger HC (1994) Induction patterns of FOS like imuno reactivity in the fore brain as predictors of atypical antipsychotic activity, *J Pharmacol Exp Ther* 271:1058–66

Rogers GA, Burke GV (1987) Neuroleptics, prolactin and osteoporosis, *Am J Psychiatry* 144:388–9.

Roth BL, Craido SC, Choudhary MS et al (1994) Binding of typical and atypical antipsychotic agents to 5-hydroxy tryptamine 6 and 5-hydroxy tryptamine 7 receptors, *J Pharmacol Exp Ther* 268:1403–10

Roth RH, Elsworth JD (1996) Biochemical pharmacology of mid-brain dopamine neurons. In: Bloom FE, Kupfer J, eds, *Psychopharmacology: The Fourth Generation of Progress* (Raven Press: New York) 227–43

Roth BL, Meltzer HY (1996) The role of serotonin in schizophrenia. In: Bloom FE, Kupfer J, eds, *Psychopharmacology: The Fourth Generation of Progress* (Raven Press: New York) 1215–27

Roth BL, Tandra S, Burgess LH et al (1995) D4 dopamine receptor binding affinity does not distinguish between typical and atypical antipsychotic drugs, *Psychopharmacology* 120:355–68

Sanders-Bush E, Canton H (1996) Serotonin receptors: signal transduction pathways. In: Bloom FE, Kupfer DJ, eds, *Psychopharmacology: The Fourth Generation of Progress* (Raven Press: New York) 431–42

Sax KW, Strakowski SM, Keck Jr PE (1998) Attentional improvement following quetiapine fumarate treatment in schizophrenia, *Schizophr Res* 33:151–5

Schatzberg AF, Cole JO, Debattista C (1997) *Manual of Clinical Psychopharmacology* 3rd edn, (American Psychiatric Press: Washington DC)

Schatzberg AF, Nemeroff CB, eds (1998) *Textbook of Psychopharmacology* 2nd edn (American Psychiatric Press: Washington DC)

Schiele BC (1975) Loxapine succinate: a controlled double-blind study in chronic schizophrenia, *Dis Nerv Syst* 36(7):361–4

Schlecte JA (1995) Clinical impact of hyperprolactinaemia, *Baillieres Clin Endocrinol Metab* 9:359–66

Schlecte JA, Sherman B, Martin R (1992)

Bone density in amenorrheic women with and without hyperprolactinaemia, *J Clin Endocrinol Metab* 75:690–1.

Schmidt AW, Lebel LA, Johnson CG et al (1998) The novel antipsychotic ziprasidone has a unique human receptor binding profile compared with other agents. Presented at Society for Neuroscience 1998

Schooler NR (1994) Negative symptoms in schizophrenia: assessment of the effects of risperidone, *J Clin Psychiatry* 55(Suppl 5):22–8

Schooler NR (1998) Comments on article by Tran and colleagues, 'Double blind comparison of olanzapine versus risperidone in treatment of schizophrenia and other psychotic disorders', *J Clin Psychopharmacol* 18(2):174–5

Schulz SC, Findling RL, Friedman L et al (1998) Treatment and outcomes in adolescents with schizophrenia, *J Clin Psychiatry* 59(Suppl 1):50–56

Sciolla A, Jeste DV (1998) Use of anti-psychotics in the elderly, *Int J Psychiatry Clin Pract* 2(Suppl 1):S27–S36

Seeger TF, Seymour PA, Schmidt AW et al (1995) Ziprasidone (CP-88, 059): a new antipsychotic with combined dopamine and serotonin receptor antagonist activity, *J Pharmacol Exp Ther* 275:101–13

Seeman P (1995) Dopamine receptors in psychosis, *Sci Am Sci Med* 2(5):28–37

Seeman P (1996) Dopamine receptors: clinical correlates. In: Bloom FE, Kupfer J, eds, *Psychopharmacology: The Fourth Generation of Progress* (Raven Press: New York) 295–302

Seeman P, Corbett R, VanPol HHM (1997) Atypical neuroleptics have no affinity for dopamine D_2 receptors or are selective for D4 receptors, *Neuropsychopharmacology* 16:93–110

Segraves RT (1985) Psychiatric drugs and orgasm in the human female, *J Psychosomatic Obst Gynaecol* 4:125–8

Segraves RT (1989). Effects of psychotropic drugs on human erection and ejaculation, *Arch Gen Psychiatry* 46:275–84

Selkin J (1979) Loxitane-C (loxapine succinate) oral liquid concentrate in the rapid tranquilization of acutely disturbed schizophrenic patients, *Curr Ther Res* 26(6):908–19

Selman FB, McClure RF, Helwig H (1976) Loxapine succinate: a double blind comparison with haloperidol and placebo in acute schizophrenics, *Curr Ther Res* 19(6):645–52

Serban G (1979) Loxapine in acute schizophrenic disorder, *Curr Ther Res* 25(1):139–43

Serper MR, Chou JCY (1997) Novel neuroleptics improve attentional functioning in schizophrenic patients: ziprasidone and aripiprazole, *CNS Spectrums* 2(8):56–64

Sharma T, Mockler D (1998) The cognitive efficacy of atypical antipsychotics in schizophrenia, *J Clin Psychopharmacol* 18(2)(Suppl 1):12S–19S

Sheitman B, Chakos M, Robinson D et al (1998) The development of treatment resistance in patients with schizophrenia: a clinical and pathophysiological perspective, *J Clin Psychopharmacol* 18 (Suppl 1): 20S–28S

Shelton R, Tollefson G, Tohen M et al (1998) The study of olanzapine plus fluxoetine in treatment-resistant major depressive disorder without psychotic features. 38th Annual New Clinical Drug Evaluation Unit Meeting, Boca Raton, Florida, 10–13 June 1998

Shih JC, Chen K J-S, Gallaher TK (1996) Molecular biology of serotonin receptors: a basis for understanding and addressing brain function. In: Bloom FE, Kupfer DJ, eds, *Psychopharmacology:*

The Fourth Generation of Progress (Raven Press: New York) 407–14

Simpson GM, Cooper TB, Lee H, Young MA (1978) Clinical and plasma level characteristics of intramuscular and oral loxapine, *Psychopharmacology* 56:225–32

Singh AN, Barlas C, Singh S et al (1996) A neurochemical basis for the antipsychotic activity of loxapine: interactions with dopamine D1, D2, D4 and serotonin 5HT2 receptor subtypes, *J Psychiatry Neurosci* 21:29–35

Small JG, Hirsch SR, Arvanitis LA et al and the Seroquel Study Group (1997) Quetiapine in patients with schizophrenia. A high- and low-dose double-blind comparison with placebo, *Arch Gen Psychiatry* 54:549–57

Smith D, Pantelis C, McGrath J et al (1997) Ocular abnormalities in chronic schizophrenia: clinical implications, *Aust NZ J Psychiatry* 31:252–6

Sowers ME, Corton G, Shapiro B et al (1993) Changes in bone density with lactation, *J Am Med Assoc* 269:3130–5.

Sprouse JS, Rollema H, Lu Y et al (1998a) Ziprasidone is a 5HT1A receptor agonist: in vivo evidence. Presented at the XXIst CINP, Glasgow, UK, 12–16 July 1998

Sprouse JS, Rollema H, Lu Y et al (1998b) In vivo evidence of central $5HT_{1A}$ agonist activity of the novel antipsychotic ziprasidone, *Eur Psychiatry* 13(4):302S–303S

Stahl SM (1996) *Essential Psychopharmacology* (Cambridge University Press: New York)

Stahl SM (1997) Awakening from schizophrenia: intramolecular polypharmacy and the atypical antipsychotics, *J Clin Psychiatry* 58:381–2

Stahl SM (1998a) What makes an antipsychotic atypical? *J Clin Psychiatry* 59(8):2–3

Stahl SM (1998b) How to appease the appetite of psychotropic drugs, *J Clin Psychiatry* 59(10):500–1

Stahl SM (1998c) Neuropharmacology of obesity: my receptors made me eat it, *J Clin Psychiatry* 59(9):447–8

Stahl SM (in press) Selecting an atypical antipsychotic by combining clinical experience with guidelines from clinical trials, *J Clin Psychiatry*.

Stockton ME, Rasmussen K (1996) Electro-physiological effects of olanzapine, a novel atypical antipsychotic, on A9 and A10 dopamine neurons, *Neuropsychopharmacology* 14(2):97–104

Street J, Mitan S, Tamura R et al (1998) Olanzapine in the treatment of psychosis and behavioral disturbances associated with Alzheimer's disease. 3rd Congress of European Federation of Neurological Societies, Seville, Spain, 19–25 September 1998

Street JS, Tamura RN, Sanger TM, Tollefson GD (1996) A comparison of the incidence of long-term treatment-emergent dyskinetic symptoms in patients treated with olanzapine and haloperidol. NCDEU Meeting, 1996

Sunderland T (1996) Treatment of the elderly suffering from psychosis and dementia, *J Clin Psychiatry* 57(Suppl 9):53–6

Swift RH, Harrigan EP, van Kammen DP (1998a) A comparison of fixed dose intramuscular (IM) ziprasidone with flexible dose haloperidol, *Eur Psychiatry* 13(4):304s

Swift RH, Harrigan EP, van Kammen DP (1998b) Validation of the behavioral activity rating scale (BARS): a novel measure of activity in agitated patients, *Eur Psychiatry* 13(4):292s

Tandon R, Harrigan E, Zorn SH (1997) Ziprasidone: a novel antipsychotic with unique pharmacology and therapeutic potential, *J Serot Res* 4:159–77

Taylor D, Drummond S, Pendlebury J (1998) Olanzapine in practice, *Psychiatr Bull* 22:552–4

Tecott LH, Sun LM, Akana SF et al (1995) Eating disorder and epilepsy in mice lacking 5-HT$_{2C}$ serotonin receptors, *Nature* 374(6):542–6

Tensfeldt TG, Wilner KD, Baris B et al (1997) Steady-state pharmacokinetics of ziprasidone in healthy elderly and young volunteers. Presented at 150th Annual American Psychiatric Association Meeting, New York, 1997

Thomas JL (1979) Loxapine oral liquid concentrate in the treatment of young adult patients with acute schizophrenic symptoms, *Curr Ther Res* 25(3):371–7

Tohen M, Zarate Jr CA (1998) Anti-psychotic agents and bipolar disorder, *J Clin Psychiatry* 59(Suppl 1):38–49

Tollefson GD, Beasley Jr CM, Tamura RN et al (1997a) Blind, controlled, long-term study of the comparative incidence of treatment-emergent tardive dyskinesia with olanzapine or haloperidol, *Am J Psychiatry* 154(9):1248–54

Tollefson GD, Beasley Jr. CM, Tran PV et al (1997b) Olanzapine versus haloperidol in the treatment of schizophrenia and schizoaffective and schizophreniform disorders: results of an international collaborative trial, *Am J Psychiatry* 154(4):457–65

Tollefson GD, Sanger TM, Beasley CM, Tran PV (1998a) A double-blind, controlled comparison of the novel antipsychotic olanzapine versus haloperidol or placebo on anxious and depressive symptoms accompanying schizophrenia, *Biol Psychiatry* 43:803–10

Tollefson GD, Sanger TM, Lu Y, Thieme ME (1998b) Depressive signs and symptoms in schizophrenia: a prospective blinded trial of olanzapine and haloperidol, *Arch Gen Psychiatry* 55(3):250–58

Tollefson GD, Tran PV (1998a) Reply to comments by Kasper and Kufferle on article by Tran and associates, 'Double blind comparison of olanzapine versus risperidone in treatment of schizophrenia and other psychotic disorders', *J Clin Psychopharmacol* 18(4):354–5

Tollefson GD, Tran PV (1998b) Reply to comments by Gheuens and Grebb on article by Tran and colleagues, 'Double blind comparison of olanzapine versus risperidone in treatment of schizophrenia and other psychotic disorders', *J Clin Psychopharmacol* 18(2):177–8

Tollefson GD, Tran PV (1998c) Reply to comments by Schooler NR on article by Tran and colleagues, 'Double blind comparison of olanzapine versus risperidone in treatment of schizophrenia and other psychotic disorders', *J Clin Psychopharmacol* 18(2):175–6

Tracy JI, Monaco CA, Abraham G et al (1998) Relation of serum anti-cholinergicity to cognitive status in schizophrenia patients taking clozapine or risperidone, *J Clin Psychiatry* 59(4):184–8

Tran PV, Dellva MA, Tolleson GD et al (1997a) Extrapyramidal symptoms and tolerability of olazapine versus haloperidol in the acute treatment of schizophrenia, *J Clin Psychiatry* 58:205–11

Tran PV, Hamilton SH, Kuntz AJ et al (1997b) Double blind comparison of olanzapine versus risperidone in the treatment of schizophrenia and other psychotic disorders, *J Clin Psychopharmacol* 17(5):407–18

Trichard C, Paillere-Martinot ML, Attar-Levy D et al (1998) Binding of antipsychotic drugs to cortical 5-HT$_{2A}$ receptors: a

PET study of chlorpromazine, clozapine, and amisulpride in schizophrenic patients, *Am J Psychiatry* 155(4):505-8

Tuason VB (1986) A comparison of parenteral loxapine and haloperidol in hostile and aggressive acutely schizophrenic patients, *J Clin Psychiatry* 47(3):126-9

Tuason VB, Escobar JI, Garvey M, Schiele B (1984) Loxapine versus chlorpromazine in paranoid schizophrenia: a double blind study, *J Clin Psychiatry* 45(4):158-63

Van der Velde CD, Kiltie H (1975) Effectiveness of loxapine succinate in acute schizophrenia: a comparative study with thiothixene, *Curr Ther Res* 17(1):1-12

Vanelle JM, Olie JP, Levy-Soussan P (1994) New antipsychotics in schizophrenia: the French experience, *Acta Psychiatr Scand*(Suppl 380): 59-63

VanKammen DP, McEvoy JP, Targum SD et al (1996) A randomized control dose ranging trial of sertindole in patients with schizophrenia, *Psychopharmacology* 124:168-75

Walsh BT (1998) *Child Psychopharmacology* (American Psychiatric Press: Washington DC)

Wan W, Ennulat DJ, Cohen BM (1995) Acute administration of typical and atypical drugs induces distinctive patterns of FOS expression in the rat forebrain, *Brain Res* 688:95-104

Wardlow SL, Bilezikian JP (1992) Editorial: hyperprolactinemia and osteopenia, *J Endocrinol Metab* 75:690-1.

Weiss JM, Kilts CD (1998) Animal models of depression and schizophrenia. In: Schatzberg AF, Nemeroff CB, eds, *Textbook of Psychopharmacology* 2nd edn (American Psychiatric Press: Washington DC) 89-133

Welch CP, Kilpatrick AT, Butler A, Tweed JA (1998) The efficacy of zotepine in reducing negative symptoms. 11th Annual ECNP Congress, Paris, France 31 October-4 November 1998

Wilner KD, Anziano RJ, Tensfeldt TG et al (1996) The effects of ziprasidone on steady-state lithium levels and renal clearance of lithium, *Eur Neuropsychopharmacol* 6(Suppl 3):38

Wilner KD, DeMattos SB, Anziano RJ et al (1997a) Lack of CYP 2D6 inhibition by ziprasidone in healthy volunteers, *Biol Psychiatry* 42:42S

Wilner KD, Hansen RA, Folger CJ, Geoffroy P (1997b) Effects of cimetidine or Maalox on ziprasidone pharmacokinetics, *Biol Psychiatry* 42:42S

Windgassen K, Wesselmann U, Monking HS (1996) Galactorrhea and hyperprolactinemia in schizophrenic patients on neuroleptics: frequency and etiology, *Neuropsychobiology* 33:142-6

Wirshing DA, Marder SR, Goldsten D, Wirshing WC (1997) Novel antipsychotics: comparison of weight gain liabilities. 36th Annual Meeting of the American College of Neuropsychopharmacology, Kamuela, Hawaii, 8-12 December 1997

Zarate CA, Narendran R, Tohen M et al (1998) Clinical predictors of acute response with olanzapine in psychotic mood disorders, *J Clin Psychiatry* 59(1):24-8

Zayas EM, Grossberg GT (1998) The treatment of psychosis in late life, *J Clin Psychiatry* 59(Suppl 1):5-12

Zimbroff DL, Kane JM, Tamminga CA et al and the Sertindole Study Group (1997) Controlled, dose-response study of sertindole and haloperidol in the treatment of schizophrenia, *Am J Psychiatry* 154(6):782-91

Zisook S, Devaul R, Jaffe K, Click Jr M (1978) Loxapine succinate (loxitane) in the outpatient treatment of acutely ill schizophrenic patients, *Curr Ther Res* 24(4):415–26

Zorn SH, Lebel LA, Schmidt AW et al (1998) Pharmacological and neurochemical studies with the new antipsychotic ziprasidone. In: Palomo T, Beninger R, Archer T, eds, *Interactive Monoaminergic Basis of Brain Disorders: Vol 4 Dopamine Disease States* (Editorial Sintesis: Madrid) 377–94

INDEX